D0519680

THE CANADA COMPANY
AND THE HURON TRACT, 1826-1853

THE CANADA COMPANY
AND THE HURON TRACT, 1826-1853

Personalities, Profits and Politics

ROBERT C. LEE

NATURAL HERITAGE BOOKS
TORONTO

Copyright © 2004 Robert C. Lee

All rights reserved. No portion of this book, with the exception of brief extracts for the purpose of literary or scholarly review, may be reproduced in any form without the permission of the publisher.

Published by Natural Heritage / Natural History Inc.
P.O. Box 95, Station O, Toronto, Ontario M4A 2M8
www.naturalheritagebooks.com

Library and Archives Canada Cataloguing in Publication

Lee, Robert C. (Robert Charles), 1937-
    The Canada Company and the Huron Tract, 1826-1853 : personalities, profits and politics
    / Robert C. Lee.

Includes bibliographical references and index.
ISBN 1-896219-94-2

    1. Canada Company — History. 2. Huron Tract (Ont.) 3. Land settlement — Ontario,
Southwestern — History — 19th century. 4. Ontario, Southwestern — History — 19th century.
5. Ontario — History — 1791-1841. 6. Ontario — History — 1841-1867. I. Title.

FC3071.L44 2004   971.3'02   C2004-903826-5

Front cover: *Goderich Harbour.* This 1858 watercolour by W.N. Cresswell of Seaforth shows former Canada Company agent (and now Crown Lands agent) Charles Widder pointing out the sights at the Goderich Harbour to his visitors. *Courtesy of LAC C5132.*
Back cover: John Galt: *Courtesy of Archival and Special Collections, University of Guelph.*
    Thomas Mercer Jones: *Courtesy of LAC C-098835.*
    William Allan: *Courtesy of Toronto Public Library (TRL).*
    Frederick Widder: *Courtesy of the Toronto Public Library (TRL).*
Cover and text design by Sari Naworynski
Edited by Jane Gibson
Printed and bound in Canada by Hignell Book Printing, Winnipeg, Manitoba

Natural Heritage / Natural History Inc. acknowledges the financial support of the Canada Council for the Arts and the Ontario Arts Council for our publishing program. We acknowledge the support of the Government of Ontario through the Ontario Media Development Corporation's Ontario Book Initiative. We also acknowledge the financial support of the Government of Canada through the Book Publishing Industry Development Program (BPIDP) and the Association for the Export of Canadian Books.

To my parents
Isabel Beatrice (Lockhart) Lee and William Ernest Middleton Lee
and their forebears of true pioneer stock
— the Burritts, the Lockharts, the Lees, the Crabbs, and the Middletons —
whose unwavering contributions helped make Canada what it is today

TABLE OF CONTENTS

ACKNOWLEDGEMENTS

I WAS UNDER NO ILLUSION two years ago that it would be an easy task to write a book on the Canada Company based on my original Master of Arts thesis on the company. I knew from the start that considerable research and writing time would be required but was very encouraged and gratified by the very tangible offers of support from communities of interest in the former Huron Tract lands. This enabled me to delve into and expand on so many of the fascinating characters of the day and locate some visuals which had never been published before. Because I was Director of the Science and Technology Division of the Department of Foreign Affairs and International Trade in Ottawa, I had little spare time, but having so many people behind me spurred me on!

John O. Graham of Goderich, a retired communications industry executive and philanthropist (who died tragically just as this book was going to print), had started the ball rolling by offering encouragement and ensuring that this book on the Canada Company would be available in Huron County libraries and schools. Others who then added their encouragement included the Huron County Historical Society, Heritage Goderich, the Huron Business Development Corporation, the Huron County Museum and Volvo Motor Graders. Goderich Township resident Paul Carroll, retired Director of the Avon-Maitland District School Board and a former President of the Huron County Historical Society, quickly took the next steps with me. While I delved into additional research and began writing in Ottawa, he kept on top of other issues and worked diligently at locating the all-important "visuals." He also very kindly scanned the original manuscript to get me started. Without Paul's dedicated support throughout the project, I could not have possibly completed the book in such a timely and fully comprehensive fashion.

Stratford-Perth Archivist Lutzen Riedstra has also been extremely helpful and supportive. He provided a number of important visuals and advice – and his unique insights into the history of the Huron Tract were invaluable.

The following were also all very generous in providing visuals and, in many cases, locating them in the first place: the Goderich Public Library, Goderich; the Huron County Museum, Goderich; Upper Canada College Archives, Toronto; the University of Guelph (Archival and Special Collections), Guelph; the Archives of the Cathedral Church of St. James, Toronto; Trinity College, University of Toronto, Toronto; the Library and Archives Canada, Ottawa; the Toronto Public Library, Toronto (Baldwin Room); the Archives of Ontario, Toronto (Special Collections); the Guelph Public Library, Guelph, and the Seaforth and Area Museum, Seaforth.

Others loaned visuals from their personal collections. They include: Cayley Hill (Goderich), Brian and Bev Jaffray (access to portrait of Tiger Dunlop), Daphne Davidson (Goderich – whose husband, William Lizars Davidson, was a descendent of Colborne Township pioneer Daniel Lizars), Aileen Davidson Fellowes (Washington, D.C. – also a descendant of Daniel Lizars), Malcolm Campbell (Goderich), Bill Trick (Goderich Township), Nancy Williams (Toronto), Ken Cardno (Seaforth) and Kelvin Jervis (Clinton).

Others provided research regarding visuals and possible sources of documents: Lynda Jones (Mitchell Archives); Kate Jacob and Cindy Farmer (Stratford-Perth Archives); Mary Smith (St. Marys Museum); Rev. Allan Livingstone (St. George's Anglican Church, Goderich); Stewart Boden (Archives of Ontario, Toronto); Alan Walker and Tania Henley (Special Collections, Toronto Reference Library); Nancy Mallet (the Cathedral Church of St. James, Toronto); Marion Spence (Upper Canada College); Ron Walker (Aids to Navigation, Canadian Coast Guard, Parry Sound); Larry McCabe (Clerk-Administrator, Town of Goderich); Ellen Morrison (Archives and Special Collections, McLaughlin Library, University of Guelph); Bill Hughey (Archivist, Guelph Public Library, Guelph); Cathy Garrick and James Sills (Seaforth Area Museum); Elaine Sturgeon (Bayfield Archives); Helen Trompeteler (National Portrait Gallery, London, England) and Robert Watts (Chief Herald of Canada, Government House, Ottawa).

Others assisted with access to gallery and museum collections of archives, records and photographs: Pat Hamilton (Huron County Museum, Goderich); Robin Wark (Sallows Gallery, Goderich); Lynn Lafontaine (Library and Archives Canada, Ottawa).

Others provided technical support, scanning and photo correction: Kelvin James (Jervis Photo Inc., Clinton); Jeremy Allin (Huron County Museum, Goderich); Elizabeth Profit (Elizabeth's Art Gallery, Goderich).

Others helped with artist's permissions: Claudia Elliott (Parry Sound – portrait image of Tiger Dunlop); Ross Irwin (Guelph – assistance in research on the location of The Priory); Ted Turner (Goderich – research assistance on properties acquired by Thomas Mercer Jones without *apparent* Canada Company sanction).

My first cousin Eleanor (Smith) Wilson of Midland thoughtfully provided background on the Middleton family beginning with the voyage of our great-great-grandfather Charles George Middleton and our great-great-grandmother Elizabeth Wise. They left Kent, England, in 1835 for York, Upper Canada, as newlyweds at ages 21 and 20 respectively, and settled in Goderich Township that year.

To my employer, International Trade Canada (the Department of Foreign Affairs and International Trade was recently split into Foreign Affairs Canada and International Trade Canada), I also owe a particular debt of thanks for their understanding in the latter stages as I completed the manuscript.

I would be remiss if I did not make note of the guidance received from the late Dr. Donald C. Masters during preparation of my original thesis. He was an inspiration. Retired York University Professor Jim Cameron, now of Kincardine, was most obliging as I delved into the Canada Company once again. For his insights, books and suggestion for the title of my book, I am sincerely appreciative as I am for his conscientious proofreading skills. My thanks also go to Professor Roger Hall, University of Western Ontario, London, and Professor J. David Wood, York University, Toronto for their help in clarifying some niggling details with regard to government and company issues both in Great Britain and Upper Canada. I am also grateful for the sage advice of Dr. David Farr, former Dean of History, Carleton University, Ottawa, and for the various help of professors Jacob Kovalio, John Clarke and John Walsh of Carleton University, along with Christine Earl of Carleton and cartographer Anita Müller, Ottawa.

My brother Chris Lee in Vancouver and sister Kathy Anderson in Tobermory, Ontario, were most helpful in providing family background material.

Natural Heritage/Natural History Inc., my publisher, have been a sheer pleasure with whom to work. They saw the potential and with editor Jane Gibson's guidance, knowledge and experience, the book has become a reality. To Shannon MacMillan, your technical help has been greatly appreciated.

Finally, to my wife, Young-Hae, many thanks are owed for her kind and total support and encouragement in this and the many other endeavours I have embarked upon over the years! To Geoffrey, Jennifer and Stephen, our children – a big thank you is due them for their understanding as I was buried in my study for so much longer than I ever imagined. Stephen – your research help over the past summer was sincerely appreciated. May it continue to inspire you in your university years ahead.

There have been so many who helped on this project. I have tried to recognize them all. If any have been left off the list, it is by way of oversight. To them, I can only apologize.

While so many helped in so many ways, for which I am grateful, the responsibility for accuracy is mine. Any factual errors brought to my attention or to that of the publisher will be corrected in subsequent editions.

Robert C. Lee
Ottawa, June 2004

PREFACE

THE YEAR 2002 WAS A significant milestone for Goderich. It was the 175th anniversary of the founding of that delightful town on the eastern shores of Lake Huron. On April 23, 1827, Scottish-born novelist and colonizer John Galt, the first superintendent or commissioner of the Canada Company in Canada, had founded Guelph and he subsequently had reported to company directors in London, England, that there was a fine site at the mouth of the Menesetung River where he planned to establish a second community. It would be called Goderich, after Lord Goderich who, as chancellor of the exchequer before his elevation to the House of Lords in the spring of 1827, had been instrumental in introducing the necessary legislation in the British House of Commons in 1826 to create the Canada Company by Royal Charter. While the directors had wanted the first town founded to be named after Lord Goderich, Galt in his inimitable fashion had called it Guelph which is an anglicized version of Guelf, the family name of the "Georges," a royal lineage for which he had great respect (the "Georges" were of the House of Hanover from Brunswick and the House of Este and Guelf – George IV was the reigning monarch from 1820 to 1830). But the naming of the towns is another story, and one that would lead to confusion and annoyance on the part of the directors for a period of six months.

Two thousand and two was a special year for me as well. It had been thirty-five years since I had written my Master of Arts thesis, "The Canada Company:

Early in 1824, a committee was organized to establish the Canada Company. Based on their success, the company's Court of Directors convened to plan next steps, including meetings with the British government. Given positive signs from the government (and prospective shareholders), company director Simon McGillivray petitioned the College of Arms, London, in December 1824, for a grant of "Arms," "Crest" and "Supporters" for the company. The "Arms" and "Crest" were granted on June 15, 1825, while the "Supporters" were granted the next day. The translation of the Latin motto reads: "The Country Does Not Alter The Race." *Courtesy of the College of Arms, London, UK, with assistance of the Canadian Heraldic Authority.*

A Study in Direction, 1826-1853," under eminent Canadian historian Dr. Donald C. Masters at the newly established Wellington College of the University of Guelph. I had also studied under Dr. Masters during my under-graduate days at Bishop's University in Lennoxville, Quebec. I had been encouraged early on to publish the manuscript by the Archives of Ontario, but thirty-five years of world travel as trade commissioner in the Canadian Foreign Service and a growing family did not leave much time to focus on doing this. It was therefore with great delight during the anniversary celebrations in Goderich that a Goderich resident, the late John O. Graham, suggested putting copies of the thesis in libraries in the Huron Tract lands with his support. That idea led to a discussion amongst Mr. Graham, Paul Carroll, Mary Ellen Jasper and myself. Why not expand it into a book? Rounding out the story was just too great an opportunity to miss given the fascinating people of the era who figured so prominently in the annals of the company, the government of the day and the settlers.

Researching and writing the original thesis in 1966-67 had been a "labour of love," not only because of my family's long association with Huron County (my forebears arrived in Goderich Township and in the Town of Goderich in the 1830s), but because I had heard so much about the Canada Company as a boy, having spent all my summers in Goderich at that stage of my life. I was always fascinated by the concept of a company being responsible for opening and settling 1.1 million acres on land which had just been purchased by government from the Chippewa First Nation. I was also intrigued about the story of the switch of the town plans of Guelph and Goderich – which turned out to be untrue. That said, I was unaware, at that stage, of the 42,000 acres of former Crown Reserves referred to as the Halton Block by the company where Guelph was established. I had merely thought that that latter territory was part of the Huron Tract as well.

While this book looks at the Canada Company from the perspective of the management and shareholders in England, the commissioners in Canada and the politics and the larger-than-life personalities that swirled around the company's formation and operation, readers may wonder why there is such a focus on the Huron Tract. After all, including the Halton Block, the company also owned 1,384,413 million acres of land in townships that had been Crown Reserves across the southern part of Upper Canada, stretching from Ottawa to Lake St. Clair and the Detroit River.

The reason quite simply is that there really isn't the sort of story there that there is with the opening and settlement of the Huron Tract. The Crown Reserve lands which the company agreed to purchase and resell basically represented one-seventh of each township surveyed before 1824. The government of the day was responsible for the operation of the townships, with Canada Company land sales made in tandem with the sales of other land. The Halton Block, given its relatively small size, was homogeneous and not a problem to administer either. The Huron Tract lands, though, were another story. They were unlike any other lands in Upper Canada in that the company had control

King George IV was the reigning monarch when the Canada Company received its charter in 1826. He was 58 when crowned in 1820 and had lived a life of debauchery. By the mid-1820s, he was overweight and addicted to alcohol and laudanum, a form of opium, and soon began to show signs of insanity, insisting he had fought at the Battle of Waterloo. He died a virtual recluse in 1830.

Courtesy of the AO S2105.

of this land, since it was turned over to them until sold. This arrangement made for an interesting scenario. What was the company's role to be vis-à-vis settlers and government? When, for example, did the company's responsibilities become the responsibility of the settlers and local government with regard to such things as the infrastructure including roads, bridges, mills, schools and the Goderich harbour? Who would play arbiter when issues/conflicts arose, and how?

Much had been written over the years about the company, with many being critical of it. They could well have been influenced by John Galt. Following his recall to London after running company operations in Upper Canada for two-and-a-half years, Galt wrote his autobiography. In it, he talked in disparaging terms about the company and its *modus operandi,* quoting out of context to suit his arguments. The other first-hand account (albeit some sixty years later) was written by Robina and Kathleen Lizars in 1896. *In the Days of the Canada Company* is their story of the first 25 years of settlement of the Huron Tract and their views of the social life of the period. Connected to the Galts through marriage (Helen Lizars, an aunt of theirs, was married to John Galt Jr.), they were not entirely objective in their account of the period in relation to the company. Commentators since then have been more objective, but were not able to see the complete story from the company perspective until the early 1960s. It was only after the company was wound up in 1953 and the company letter books subsequently became available at the Archives of the Province of Ontario that one was able to delve into the files to make that assessment.

I had that opportunity upon pursuing my graduate studies at the University of Guelph. Company letter and minute books had been microfilmed and their contents were waiting to be discovered.

In considering the topic of the company, with its operations in Canada and the settlers on the one hand, and the company directors in London on the other, it is important to put conditions of the day into context. Company directors and company commissioners were 3,000 miles apart, with the latter in the untamed wilderness of Upper Canada. Return mail took at least eight weeks and more likely ten. There had to be mutual trust and understanding on both sides. Through its commissioners who ran the operation in Canada, the company had to try to maintain a "middle-of-the-road" policy between government officials and settlers – and doing so was never easy. Furthermore, if the commissioners in Canada had their unique challenges, so did the directors in England during the early period of the company. Galt was altruistic but failed to understand what was going on back home. Shareholders were skittish and share value was dropping, while Galt's reporting (financial and otherwise) left much to be desired. He might have been a favourite of the settlers, but his bookkeeping was inept, to say the least, and he certainly ruffled provincial government feathers. In so many words, he seemed to see himself as next to the lieutenant governor in importance. Shareholder support was being lost because the directors could not provide meaningful financial statements, and shareholders did not really understand the nature of their investment; they were expecting a quick return, which was quite unrealistic. The directors had no choice but to recall John Galt. By his actions, he had not only alienated the colonial government but his masters back home as well.

The new commissioners, the wealthy and well-connected William Allan of York, Upper Canada, and Thomas Mercer Jones, who was newly arrived from England, confirmed the directors' worst fears about the condition of the company's books in Upper Canada, such as they were. Allan reported very frankly on the state of affairs and offered a solution. To compound matters, there was then the cholera epidemic of the early 1830s, the conditions leading up to the Rebellion of 1837, the rebellion itself – and its aftermath.

The decade following the rebellion was not without its challenges. While new Canada Company commissioner Frederick Widder, who was appointed in 1839, more than proved his worth, Jones' *modus operandi* was certainly cause for angst amongst the directors. But if the business side of Jones' life was somewhat rocky, Jones and his wife were the doyennes of society during their time in Goderich. Was it Jones' marriage to Archdeacon (then Bishop) Strachan's daughter that kept him in the employ of the company until 1852?

While the company motto *Non Mutat Genus Solum* (the country does not alter the race) points out that the settlers should still feel British, many new arrivals felt the company had not done enough for them. Consequently, the Canada Company came in for a great deal of criticism, much of it unjustified in light of circumstances. One only has to read the *Illustrated Historical Atlas of the County of Perth*, which was published in 1879, to see the spiteful language used:

> The ... evidence ... both from the oldest settlers and all disin-
> terested publications ... goes to prove conclusively that the
> "Canada Company" were, through and through, the most
> unconscionable and unscrupulous ring of "land-grabbers"
> which this country, ... has any knowledge of.[1]

But not all authors have been so harsh in commenting on the company. Perhaps William Johnston in the *History of the County of Perth*, published in 1903, was more objective:

> It may fairly be said that a man's want of success in his business
> affairs will in almost all cases be attributed to every known or
> conceivable cause except the correct one. In his endeavour to
> satisfy his feelings he will never accuse himself as being the
> cause of his own misfortunes. The vagaries of luck, combina-
> tions of circumstances, perfidious friends, commercial exigen-
> cies, duplicity of those with whom he has business relations, are
> the spirits of evil that has crossed his path, but never himself,
> who may be the worst spirit of them all.
>
> From a lively exercise of this principle arose much of that
> discontent regarding those methods by the Company for set-
> tling their lands.[2]

The Canada Company was founded at a time when no consistent system of management and settlement existed in Upper Canada. It was born directly because of this condition, coupled with the ramifications of the Constitutional Act of 1791, the Napoleonic Wars and the losses suffered by inhabitants of the Niagara frontier emanating from the War of 1812-14. Inevitably, when a public stock company acquires close to two and one-half million acres of land, controversy is bound to arise among the key players: settler, government and company. It did.

In summary, the personalities of the day in the first half of the 19th century in England and Upper Canada were larger than life. These included politicians in both countries, company directors and commissioners, church leaders and settlers. Each had a point of view which did not necessarily mesh with the other. Superimposed on all of this were the company shareholders who expected to make a quick profit – but didn't, at least in the short term. These groups made for interesting, if not strange, bedfellows.

Throughout the book there are a number of quotes taken directly from either company records or other resources. In the interest of historical authenticity, these quotes are shown in their original form.

It has been said that when writers of history undertake their work with a specific goal in mind, they are bound to be selective in their facts. Instead of letting the record speak for itself, they emphasize those aspects which support their thesis and underplay others. I hope that in this treatment of the Canada Company, the reader will agree that a fair interpretation of the facts has been presented. Furthermore, it is my hope that the research which went into the original thesis – and my subsequent findings over the past two years – will help those who are particularly interested in the settlement of the Huron Tract to understand what went on behind the scenes. These machinations lead not only to the founding and operation of the company, but also to the settlement of this most attractive part of Ontario while the history of Canada was in the making.

Robert C. Lee
Ottawa, June 2004

A NOTE FROM THE AUTHOR:
MONEY IN CANADA 1763-1858

When the French colony of Quebec became a British possession with the signing of the Treaty of Paris in February 1763, pounds, shillings and pence became the official money of account in Canada and remained so until 1858. In practice, this meant that colonial governments set a value for the various kinds of money circulating, including Spanish dollars, American dollars, American gold coins and "army bills," which were used by the British army during the War of 1812-14. The physical monies were rated as coins in the money of account, the value for which changed from time to time. If it was overrated, the money would flow in in large numbers while, if underrated, it would flow out to wherever it was rated higher.

The money of account most commonly used in Upper Canada during the period covered by this publication was the Halifax (or Canadian) currency – worth approximately ten per cent less than sterling (and commonly referred to as "currency"). York currency was also used, but to a much lesser extent. It was equivalent to three-quarters of Halifax currency and was outlawed in accounts with the government in the 1820s although usage continued in some areas of the province for another 20 years. While both coins and paper money (to a limited degree) were in circulation pre-war, Canadians only grew accustomed to using bills as reliable money because of the circulation of British "army bills" during the War of 1812-14.

Dollars (and its parts) were the most commonly used physical money in

Upper and Lower Canada and it became common practice to use them as money of account in private transactions. However, the colonial government and British companies like the Canada Company would have kept accounts for the most part in official Halifax currency. But to confuse things further, the company used a combination of Halifax currency and pounds sterling to pay their officers. For example, salaries were paid quarterly in pounds sterling while allowances were paid in "currency."

During the time period of this book, one pound Halifax or Canadian "currency" generally equalled $4.00 Canadian which was equal to $4.00 U.S. One pound sterling equalled $4.44 Canadian. Unless otherwise noted, the currency noted is "sterling."

In the late 1840s, the government of the Province of Canada opted to drop pounds, shillings and pence and adopt a decimal system but the change over was gradual. In 1858, the Canadian government then legislated that government accounts were to be kept in dollars only rather than in pounds, and began to issue and circulate its own decimal money along with that of the banks.

With reference to Canada Company land values in Canada, negotiations with the British government would have undoubtedly been in pounds sterling because of having taken place in England. However, the land values that the Canada Company charged settlers in the Huron Tract would have been in pounds currency. This will explain the cost of the lands commonly stated as 3s6d currency and 3s2d sterling[1]

ABBREVIATIONS

AO	Archives of Ontario
B & LHR	Buffalo and Lake Huron Railway
BB & GRC	Buffalo, Brantford & Goderich Railway Company
CPR	Canadian Pacific Railway
CT & LHRRC	City of Toronto and Lake Huron Rail Road Company
GJR	Guelph Junction Railway
HBC	Hudson's Bay Company
LAC	Library and Archives Canada
MP	Member of Parliament
MPP	Member of Provincial Parliament
T & GRC	Toronto and Guelph Railroad Company
TRL	Toronto Reference Library

CHAPTER 1

WHY THE CANADA COMPANY?

*Whether there is anything to regret ... no man can tell — that
there is no use in regretting it — is very certain.*[1]

W ITH THIS COMMENT, the Chief Justice of Canada West, John Beverley
Robinson, reflected on the formation of the Canada Company some
twenty-nine years earlier in 1824. Born of Loyalist stock in Lower Canada in
1791, and schooled firstly in Kingston and then in Cornwall by the young Scot,
John Strachan, who would later become very influential in the colony and with
whom Robinson would develop a personal lifelong friendship, Robinson had
been attorney-general of Upper Canada during the six-year period that the
concept of establishing the company was being explored and promoted. He had
been concerned with the plight of the colony over the vexing issue of the
Clergy Reserves[2] and the Crown Reserves — areas of land representing two-
sevenths of each township which had been set aside equally, one-seventh for
government use and one-seventh for the support of the clergy in Upper
Canada to counterbalance a similar previously granted privilege to the Roman
Catholic Church in Lower Canada.

The church rights controversy aside, the more significant issue was that
these lands were generally standing idle and inhibiting settlement. To date, no
consistent policy had been followed as to their disposition. To this end, Robinson

As the young attorney general of Upper Canada, John Beverley Robinson was concerned about the financial plight of the colonies and the adverse impact the Clergy and Crown reserves were having on the settlement of the province. He claimed that the concept of a company purchasing these lands to pay colonial expenses and bringing out settlers was his idea. *Courtesy of LAC C111481.*

had presented a paper on the subject in 1823 in London, England, while study-ing law at Lincoln's Inn in the British capital.

The Constitutional Act of 1791 which had in part created the reserves pro-vided Upper and Lower Canada with the structure of Governor, Executive Council and elected Assembly. Since the issue of taxes had provoked the American Revolution, and land taxes in particular were viewed as an impedi-ment to settlement, the idea was that government in the Canadas would be financed by reserving a seventh of all land for the Crown which would be sold over time on a consistent basis to meet day-to-day expenses. To the colony, however, both of the reserves were becoming a liability. What Robinson said about the situation in Upper Canada was not new, but it had the effect of crys-tallizing the grievances. He felt so strongly on the subject, in fact, that he had not only presented his views to the Lieutenant Governor of Upper Canada, Sir Peregrine Maitland,[3] but he had also had private discussions in London, England, with his close friend, the Undersecretary of State for the Colonies Robert Wilmot-Horton.[4] In this position as undersecretary, Wilmot-Horton was able to play a key role since Lord Bathurst,[5] the Secretary of State for the Colonies, left much of the daily routine of running the office to him, and it is interesting to note that men like Robinson were often co-opted to assist the Colonial Office in formulating policy involving the colonies. It also helped that Robinson had been introduced to Lord Bathurst and that Wilmot-Horton and Robinson had become close friends while Robinson was studying in London. Robinson's friendship with Strachan would also be an important factor as he, Strachan and Wilmot-Horton worked through how to best solve the financial issues facing Upper Canada and to encourage emigration there.

John Galt,[6] the Scottish-born novelist, poet, biographer, historian, essayist, critic, lobbyist and sometime businessman, had been working in another area of colonial grievances, specifically the property losses resulting from the War of 1812–14. He had been retained by Loyalist interests to press their claims against the British government for war reparations. He, too, had been in conversation with Wilmot-Horton over financial payments for those suffering losses.

The Canada Company was born directly from these two seemingly unrelated colonial grievances. The company did not just "happen," but was the result of intensive work and perseverance on the part of a group of London businessmen, certain colonial officials and John Galt, who was determined to push the concept through to fruition.

Even after the idea of establishing the company was accepted, and a Provisional Committee established to make it happen, two years would elapse before a charter would be granted. In the meantime, shareholder support would become uncertain, and much financial support lost. Yet, in spite of the many uncertainties, the company not only managed to survive, but company stock became sought after by the mid-1830s.

Conditions Leading to the Formation

It is difficult to say who first conceived of the idea of the Canada Company. If John Galt is to be believed, it was he who first suggested the systematic sale of the reserves, and the formation of the company. If Robinson's account is accurate, he was the one who had suggested it first. According to Robinson, Galt had merely talked to Wilmot-Horton, and had read Robinson's paper on the subject. He, then:

Sketch of the manner in which a Township is laid out in Upper Canada

Crown Reserves . . . . +
Clergy Reserves . . . . O
Perpendicular Double Lines are the Public Roads
Horizontal Double Lines are Concession Roads

Detail from a large map of Upper Canada printed by the Canada Company prior to 1828 showing the arrangement of reserved lots in older townships across the province. Their sale supported the government and the Anglican church.

Plan of a typical township in Upper Canada. The layout of the townships, with one-seventh reserved for the Crown, and one-seventh for the Clergy, inhibited settlement. *From the author's collection.*

> … being a shrewd scheming man … immediately conceived
> the idea of getting up a company & making a grand speculation
> of these Crown Reserves of which he knew as little as Mr.
> Horton did before I gave him my paper.[7]

The Clergy Reserves issue was most divisive given the prominence of the Church of England and the pressure being exerted by other Protestant groups for their fair share of any monies realized from their sale. The Crown Reserves issue, on the other hand, was less divisive, the lands having been set aside for "public purposes." In Upper Canada, because sales had been hindered by the lack of a consistent policy relating to them, settlement could not spread in an orderly fashion. Furthermore, the scarcity of adequate roads in the area of the reserves meant that communication and travel were hampered.[8]

The reserve lots of two hundred acres each had been distributed at more or less equal intervals in the townships that existed in 1791 (because the Niagara District had been settled before 1791, a large 42,000-acre block of land had been set aside in what is now Guelph Township of Wellington County, which the company referred to as the "Halton Block).". This piece of land was to compensate for a lack of Crown Reserves in Lincoln County. The practice meant that of the 66,000 acres which made up each of most townships, 18,850 acres had been set aside for the Crown and Clergy reserves. Well before any real grievance over the Clergy Reserves was lodged on religious grounds, general discontent was brewing over the obstruction to settlement which they posed. Because the sale of these reserves to settlers was not being actively promoted, there was no consistent pattern of settlement in the townships. Furthermore, roads were not being properly maintained in front of the reserves.

While half-hearted attempts had been made to *lease* the Crown Reserves, these had been unsuccessful as many settlers were hesitant to clear and settle leased lands. Furthermore, while all lots, in theory, were leased at a uniform rate regardless of location, and many could be obtained from the Crown "almost gratuitously," others were offered rent free.[9] Since 1802, approximately 1,200 lots had been leased, but many had been taken up by speculators who had no intention of either residing on them or cultivating them. What rents were being paid were being remitted only irregularly. Instead of trying to collect back monies due that year, the authorities decided to raise rents in an effort to improve the revenue stream. This move still did not work in the government's favour. Furthermore, no distinction regarding the amount of rent to be paid had been made as to whether the property was near a settlement, on a river, or off in the wilderness.

Working capital was far from abundant in Upper Canada, and Britain hesitated to invest any more than was absolutely necessary in the colony. Not only had the War of 1812–14 been costly, but twenty-two years of war on the continent had cost Great Britain dearly. For example, in just three years between 1813 and 1815, the British government had poured £32,000,000 in subsidies into Europe to support the fight against Napoleon and to deal with its aftermath.[10] Furthermore, the British Corn Laws had directly affected the colony's market in Great Britain since neither wheat nor flour could be sold in England if the average price of wheat was below sixty-seven shillings sterling per quarter (eight bushels).[11] As Upper Canada had very little industry at this stage, it depended almost exclusively on agricultural production, yet the new legislation had effectively cut off the British market for the colony as the current price for wheat and flour was below the prescribed minimum.[12]

To add further hardship, the colony had been paying at least two-thirds of the cost of running the colonial government (more than £20,000 annually). Furthermore, they had been forced to borrow funds for their day-to-day operations because of a dispute with Lower Canada which meant they no longer had access to what was judged by them to be a fair share of customs receipts.[13] In all, there was barely enough revenue to cover expenses; financially, the colony was in trouble. A land tax, which was an obvious source of revenue, was judged to be out of the question, (as noted earlier), since such a tax would most certainly have an adverse effect on settlement.[14] As well, Upper Canada had already spent more than £60,000 on pensions for militia officers, wounded men and families of war dead from the War of 1812–14 and was now faced with paying some three to four thousand pounds yearly for these pensions.[15]

A proposed loan to aid those who had suffered property damage during the war brought the colony's financial crisis to a head. While legislation had been prepared in Britain to authorize the British government to join Upper Canada in raising £182,130, this would have entailed a further expenditure for the colony of between five and six thousand pounds in interest alone, without taking into account repayment of the principal.[16] A solution had to be found. John Galt and John Beverley Robinson both thought they had the answer.

Since the late eighteenth century, the British government had supported, and attempted, a number of approaches to the settlement of lands in Upper

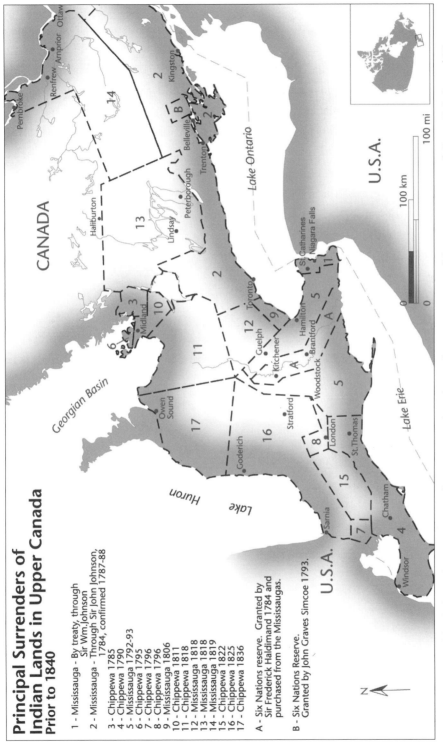

Principal Surrenders of Indian Lands in Upper Canada
Prior to 1840

1 - Mississauga - By treaty, through Sir Wm. Johnson
2 - Mississauga - Through Sir John Johnson, 1784, confirmed 1787-88

3 - Chippewa 1785
4 - Chippewa 1790
5 - Mississauga 1792-93
6 - Chippewa 1795
7 - Chippewa 1796
8 - Chippewa 1796
9 - Mississauga 1806
10 - Chippewa 1811
11 - Chippewa 1818
12 - Mississauga 1818
13 - Mississauga 1818
14 - Mississauga 1819
15 - Chippewa 1822
16 - Chippewa 1825
17 - Chippewa 1836

A - Six Nations reserve. Granted by Sir Frederick Haldimand 1784 and purchased from the Mississaugas.

B - Six Nations Reserve. Granted by John Graves Simcoe 1793.

The surrender of lands from the Native Peoples was crucial to the settlement strategy of the governments of Great Britain and Upper Canada. Not shown are the Chippewa Reserves of Kettle and Stony Point on Lake Huron. *Map created by Anita Müller.*

Canada. The "Leader and Associates" system had been tried with mixed results. Under this scheme, an individual took responsibility for the settlement of a new area with certain associates. Various other schemes had also been put forward to assist or subsidize particular groups to leave the British Isles and settle in the colonies, including large groups of Scots.[17] While Britain's entry into the Napoleonic Wars had checked emigration, the coming of peace after the Battle of Waterloo (1815) had created grave economic problems in Britain. Thousands were unemployed and on relief. Furthermore, the enclosure movement in the Lowlands of Scotland and the clearances in the Highlands had displaced many from the land because the development of pasture lands for sheep farming and the grazing of cattle had replaced traditional crofters and tenant farmers. In Ireland, on the other hand, the population was growing while the peasant's standard of living was steadily deteriorating.[18]

Of the various schemes put forward as solutions to the problems facing Upper Canada, John Beverley Robinson's plan for the disposal of the Crown Reserves would appear to have been the first suggested by an important colonial official.[19] The Colonial Office was prepared to listen. Robinson's proposal had a dual purpose.[20] Not only would it promote orderly settlement, but it would prevent the Crown from being at the mercy of the House of Assembly for the ordinary expenses of government. Robinson was anxious to find monies for government operations "in a manner least bothersome to the Province."[21] The so-called "Family Compact" (a disparaging term for the inner circle of prominent colonial members of the Judiciary, Executive and Legislative councils and senior bureaucrats) was suspicious of the Assembly who wanted more power. If the sale of the reserves and other revenue controlled by the Executive Council would meet the cost of running Upper Canada, so much the better. Robinson asserted:

> ... such a course of things would by running the most fruitful course of difficulties and unpleasant discussions between the Government and the Legislature, secure that harmony between them, which is so necessary to the prosperity of the country.[22]

John Galt had shown an interest in Robinson's proposals during his role as agent for the war-loss victims in Upper Canada. In July 1816, a Board of

Novelist, adventurer and lobbyist, John Galt said the Canada Company was his idea. To this end, he actively and successfully promoted its establishment. Sadly, he was his own worst enemy, a trait which ultimately did him in. A "big picture person," he lacked management and diplomatic skills. Company directors terminated him in January 1829. *Courtesy of Archival and Special Collections, McLaughlin Library, University of Guelph, H.B. Timothy Collection.*

Inquiry, created at the behest of Colonial Undersecretary Wilmot-Horton, had been prepared to grant a sum of £230,000 indemnity to these claimants. However, because the British government considered the amount too generous, they appointed a Commission of Enquiry which in turn reduced the award to £182,130. The delay in granting reparations annoyed the victims to the point that they had retained Galt to look after their interests to press their claims. Galt was representing such individuals as Colonel Thomas Clark, a prominent Niagara District businessman, militia officer and politician and one of the commissioners appointed for assessing war losses in the Niagara District;[23] Upper Canadian businessman Robert Grant from the Niagara region; Robert Nichol, a lawyer and quartermaster-general in the militia during the war who became a commissioner of roads and later a judge of the Surrogate Court for the Niagara District;[24] and William Dickson,[25] a cousin of John Galt. Dickson was a prominent lawyer, businessman, politician and land speculator/colonizer. In 1815, he became a member of the Legislative Council and that same year he purchased a total of 90,000 acres along both sides of the Grand River which became the townships of North Dumfries (in present-day Waterloo Region) and South Dumfries (in Brant County).

A commission of 3% on the value of recovered claims had been promised to Galt by his clients. As he was in difficult financial straits at the time, with a wife

John Galt was born in the middle flat of this tenement building in Irvine, Scotland, in 1779, to his seafaring father, Captain John Galt, and mother, Jean Thomson. Sickly as a child, he was often schooled at home. When he was 10, his family moved to Greenock. Galt entered the workforce as a customs clerk there at age 16. *Courtesy of Archival and Special Collections, McLaughlin Library, University of Guelph, H.B. Timothy Collection.*

and three young children to support, this was a welcome assignment[26] It was in this capacity that Galt had been in touch with Robert Wilmot-Horton and had seen John Beverley Robinson's paper on the subject of the reserves.[27]

The Formation

The period of the formation of the Canada Company is a fascinating one. The behind-the-scenes manoeuvres among Galt, Wilmot-Horton, Bathurst, Strachan and John Beverley Robinson and the Provisional Committee of the Canada Company are intriguing. The issues involved not only a company taking over land for settlement in Upper Canada, but *what* land and for *what* price. Adding a further complication were the Clergy Reserves. Galt had been pressing for the acquisition of these lands by the proposed Canada Company, while the now-prominent Anglican cleric and Executive Council member Archdeacon John Strachan[28] would have no part of Galt's plan as presented.

John Strachan was born in 1778 in Aberdeen, Scotland, and had arrived in Upper Canada in 1799 fresh out of university. During his early days of teaching the up-and-coming generation of young men in Kingston and then Cornwall, he was ordained a deacon in the Church of England in 1803 and became a government-paid missionary in addition to his teaching duties. In 1807, he married Ann Wood, the widow of fur trader Andrew McGill. In 1812, he and his wife

Archdeacon John Strachan as a young man. Born in Aberdeen, Scotland, in 1778, John Strachan became an important figure in Upper Canada due to his connections (as a young educator he tutored the sons of the elite, who later moved into positions of authority) and as a member of the Family Compact. He was a man of tenacity, wit and intellect, but ever the "Tory." He contributed much to education and founded Trinity College in Toronto in 1851. He died in 1867. *Courtesy of the University of Trinity College Archives.*

The residence of Archdeacon John Strachan was built in 1818 on Front Street between York and Simcoe streets. The Archdeacon (later Bishop) and his family lived well but were always short of money. Strachan arrived in Canada as a 21-year-old teacher in 1799. He became a priest in 1804 and married Ann Wood McGill of Montreal in 1807. They had eight children, four of whom lived to adulthood. *Courtesy of the Toronto Public Library (TRL), J. Ross Robertson, Collection, JRR 685.*

LAKE HURON
A Map of the
Province
of
Upper Canada
North America
Compiled by James G. Chewett.
Thomas Ridout Esq. Surveyor General
of the Province
Crown
by the
Canada Company.
Incorporated 1826.
Governor.
Deputy Governor.
Directors.
Auditors.
Secretary.
MICHIGAN TERRITORY
To
His Most Excellent Majesty
King George
IV.
This Map
LONDON
LAKE ERIE

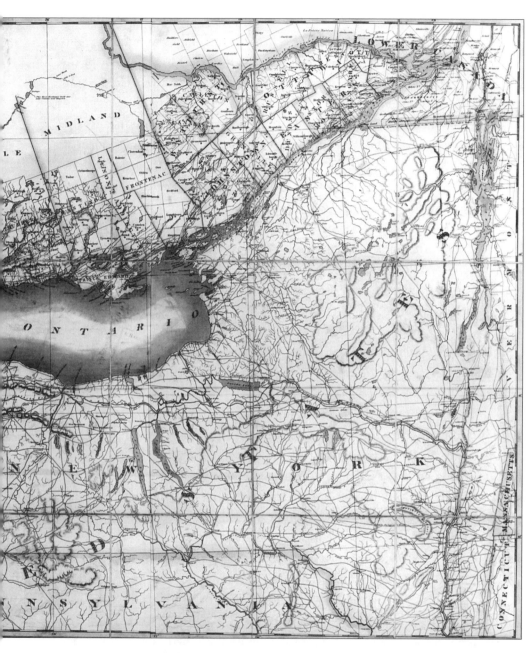

Map of the Province of Upper Canada, 1826. This map was compiled by the Canada Company when it assumed it would be purchasing approximately 1.3 million acres of Crown Reserves and some 829,000 acres of Clergy Reserves in Upper Canada. This explains why the Huron Tract lands adjacent the southern half of Lake Huron are not indicated. *Courtesy of AO F129, ATLAS 179 AO 2810.*

and young family moved to York where he showed considerable leadership during the War of 1812-14. Many of Strachan's former pupils were now beginning to move into positions of authority. Strachan was made a member of the Executive Council in 1817 and of the Legislative Council in 1820. He was now in a position of influence.

The original plan, as submitted by John Galt to Lord Bathurst, was to appoint a special agent to do "what the government is now doing at so great an expense." The agent would make arrangements for the sale of the lands with his salary based on a commission for lands sold.[29] Having changed his mind as to this approach three weeks later, Galt asked:

> would [it] be practicable to form a company to find individuals who would be willing to embark capital in the purchase from Government on a considerable scale and undertake to clear and settle lands ....[30]

This was quickly followed up with another letter the next day to Lord Bathurst in which he referred to the reserves "of which the government have resolved to dispose."[31]

The bold assumption that the government had resolved to dispose of these lands did not sit well with Bathurst. Undersecretary Robert Wilmot-Horton replied curtly on Bathurst's behalf that the government was only prepared to consider the question of changes in management of the Crown Reserves *after* appropriate information had been gathered on the subject. On the question of forming a company, he stipulated, "I am persuaded that it would be rejected as a measure to be carried into execution *immediately*."[32]

An attached memo seemed to put an end to the proposal. It read that Lord Bathurst had decided to adopt, in part, a suggestion by Sir Peregrine Maitland that an office for the sale of Crown lands was to be established with Peter Robinson,[33] the oldest brother of John Beverley Robinson, to be offered its management. Peter Robinson was entrepreneurial. He had been a fur trader, had served in the militia with the rank of captain and had successfully invested in land. He was elected to the House of Assembly where he sat from 1817 to 1824, and had chaired a committee on St. Lawrence River navigation and, in 1823, became a director of the Bank of Upper Canada. Through the good offices of his brother John, he had been introduced to Undersecretary of State

Peter Robinson was concerned about the plight
of the Irish and led two impoverished Irish
groups to Upper Canada; Peterborough is named
after him. He felt strongly that facilities such as
mills, churches and schools should be built before
settlers arrived. The Canada Company favoured
wealthy settlers who would build such facilities.
Courtesy of the Toronto Public Library (TRL), J. Ross
Robertson Collection, T15128.

Wilmot-Horton who had obtained British government approval for a small-scale experimental scheme to send poor and dispossessed Irish farmers to Upper Canada as settlers. The undersecretary had taken an immediate liking to him and, with Wilmot-Horton's support, Robinson would go on to successfully establish groups of Irish settlers in present-day Peterborough and region in 1823, and similarly in Lanark County in 1825. Sir Peregrine Maitland was also impressed with Robinson. Because Wilmot-Horton held him in such high esteem, Maitland would certainly have sanctioned the undersecretary's choice of appointing Robinson as Commissioner of Crown Lands (which, in fact, did not happen until 1828).

While the Crown land discussions were taking place, John Galt, in meantime, had written to Wilmot-Horton on the subject of the "Jesuits' property" (the Clergy Reserves).[34] The stage had been set.

If Galt actually thought that Strachan would sanction the sale of the Clergy lands to the company, he was sadly mistaken. To Strachan, these lands were sacred territory. As Peter Robinson noted:

> ... nothing could be more prejudicial to the best interests of the
> Province than that any idea ... should even be suffered to lead
> to an encroachment upon the Clergy Reserves.[35]

Archdeacon Strachan's strongly held position was that if a radical change in the management of the Clergy Reserves was contemplated, any initiative to do so

An activist and knowledgeable Roman Catholic priest, Father (later Bishop) Alexander Macdonell helped displaced Scottish Highlanders and discharged soldiers emigrate to southeastern Upper Canada. His knowledge of the colony was invaluable as Galt developed the Canada Company concept. The town of Alexandria in Ontario's Glengarry County is named after him. *Courtesy of the Toronto Public Library (TRL), J. Ross Robertson Collection, T16090.*

must lie with the Clergy Reserves Corporation which had been set up in 1819 to manage them.[36]

Undaunted, Galt suspected the government was not completely adverse to his proposal. He set about to interview important colonial personages on the subject, in particular Archdeacon Strachan, Peter Robinson and the Roman Catholic cleric Father Alexander Macdonell,[37] who had become a titular bishop in 1819 and would be appointed bishop of the Diocese of Kingston in 1826. Macdonell was a tireless settlement promoter and a central figure in the religious and political life of Upper Canada.

By way of such conversations with key players in Upper Canada, Galt would become conversant with the issues. His subsequent letters to Wilmot-Horton were persuasive. The payments for the land, he believed, would "be the first money from any colony in the English Exchequer."[38] Such a proposal was an attractive argument for the hard-pressed British government, which was struggling financially with the post-war depression that followed the Battle of Waterloo and the end of the Napoleonic Wars. He carried on his campaign with vigour:

never yet had one single capitalist come into the country, purchased large tracts of land, built mills, made roads or in any
manner prepared it for settlement.[39]

Galt compared the backward condition of Canada to the Genesee area of New
York State which had been settled by land companies. He used the fear of a
possible American take-over of Canada and the attraction of the reserves to
would-be invaders:

> Is it not known to your Lordship that the Crown and Clergy
> Reserves were held out as lures to the Americans in the late
> War? Is it impossible that occasion may arise hereafter to place
> them in the same light?[40]

Wilmot-Horton, in the meantime, had himself been collecting evidence by
interviewing key personages back in England on their travels and/or by correspondence. These people included Colonel Thomas Talbot,[41] a settlement promoter and "father" of the half a million acre Talbot settlement in southwest
Upper Canada; Archdeacon John Strachan; Father Alexander Macdonell; Peter
Robinson and William Gilkison,[42] a merchant and a ship's captain on the Great
Lakes, who was also a cousin and former schoolmate of John Galt. He enquired
into the commerce of the colony, land values and the condition of roads.

On the basis of favourable responses to the idea of establishing a land
company and the settlement possibilities it offered, Lord Bathurst was willing to
listen to Galt's plan. He would extend to the proposed company the same principles as had been granted to the Australian company which had been set up to
purchase government lands in Australia and recruit emigrants to settle on them.[43]

Two Years of Frustration

Although Lord Bathurst had been willing to sanction the formation of the
company, the two-year period of protracted negotiations and discussions
nearly resulted in the collapse of the company before it even began.

A Provisional Committee was established in England to obtain the company's
charter and make arrangements to purchase the Crown and Clergy reserves.[44]
The plan had been to send six commissioners to Canada to value the reserves,

but this was idea was scrapped when the difficulty of coming to an equitable standard value quickly was realized. Furthermore, Archdeacon Strachan was making trouble. Thus, instead of asking for all of the Clergy lands, the committee set its sights on half (of those surveyed before March 1, 1824) and all of the Crown Reserves.[45] The Provisional Committee proposed to acquire annually not less than 800 lots, or 160,000 acres of these lands, at a fixed price of £20,000 per annum for fifteen years, with an option to purchase further land at a fixed price of 2/6 an acre. This would save the expense and delay of valuation, and would allow the proposed company to organize immediately.[46]

Responding on behalf of Colonial Secretary of State Lord Bathurst, Wilmot-Horton would have none of it – his reply was straightforward:

> ... with reference to this specific proposal, Lord Bathurst has examined all the Documents in the Colonial Office ... and has also had recourse to ... oral information; ... the result ... is, that the proposition ... of the intended Canadian Company is absolutely inadmissible and that the difference between the value as estimated by the ... Company, and that which his Lordship has derived from other sources is so great as to make him less sanguine in supposing that any common principle of arrangement can be agreed upon without having recourse to the expedient of valuation by the medium of Commissioners, which though it may be attended with inconvenience in some respects, is alone calculated to reconcile the differences which exist on the subject.[47]

Calling on information provided by Dr. Strachan and Peter Robinson, Lord Bathurst noted tersely through Wilmot-Horton:

> It is understood that no land in Upper Canada unless removed from any settlement, can be bought at less than 7/6 per acre.[48]

But of course from his comfortable surroundings in London, Lord Bathurst had not taken into account the fact that most of the land was indeed far removed from settlement!

Bathurst was perfectly aware of the necessity of introducing capital into the

Lord Bathurst was secretary for war and the colonies from 1812 to 1827. A pragmatist, he was past his prime in his latter years, which meant Undersecretary Robert Wilmot-Horton (1821-28) was able to exert a fairly significant influence on colonial matters in British North America. These were crucial years for the Canada Company. *Courtesy of LAC C3838.*

colony and he was sympathetic to the company deriving a "liberal advantage" (in the event that it was ever formed). However, if the government accepted the committee's proposition as offered, he was convinced that profits would be greatly disproportional to the average profits from "similar speculations." This would be a bonus for the company at the expense of the public.[49]

Galt replied quickly. He had contemplated that the proposed company would be "unpalatable" to many connected with the local administration of the colonial government – especially to those who managed the land-granting department. That said, he could not personally envisage that it would be "unpalatable" or disadvantageous to the people or unacceptable to "His Majesty's Government" since there were no funds available to assist settlers in the first place:

> Nor are they likely to obtain from Parliament the means for the objects proposed to be undertaken by the Company without which means the existing lethargy in the colony must continue…
>
> … your Lordship is also well aware from the management of the Colonial Lands hitherto how far the proceeds arising from them have been available as a source of revenue, and can best

infer what prospect there is of the system which has existed for more than thirty years in Upper Canada, rendering in fifteen years any such returns ... as the offer for the proposed Company.[50]

In an effort to make the proposal more attractive, the British government was offered a share in the expected profits. If at any time the company was able to pay a larger dividend than the rate of interest in the colony (6%), or to pay any bonus to the stockholders, the government would be given one-third of the extra.[51]

Lord Bathurst had the final word. In a letter addressed to the Provisional Committee, dated July 24, 1824, he consented to the price being fixed by five commissioners. The land was to be evaluated according to the average price obtained for uncleared lands as of March 1, 1824,[52] in the established districts of the colony (from east to west) Eastern, Johns Town (or Johnstown), Midland, Newcastle, Home, London and Western. While justice was to be done to the public, the company was nonetheless to be protected from ultimate loss. It was to receive "that fair and ample remuneration which is due to those who embark capital in an undertaking to which the public advantage is so materially concerned."[53]

In a letter addressed to Wilmot-Horton, Canada Company director Simon McGillivray[54] pressed for an immediate granting of the charter. McGillivray, the son of a poor Scottish tenant farmer, was a businessman who had been involved in the fur trade from his teenage years. With the help of the highly successful businessman and independently wealthy Edward Ellice, MP,[55] another Canada Company director, he had been a key figure in the merger of the fur-trading North West Company and the Hudson's Bay Company in 1821. Following the merger, two companies also involved in the fur trade, in which McGillivray was a partner – McTavish, McGillivrays and Company and McGillivray, Thain and Company – went bankrupt in 1825. McGillvray was now in a position to devote his not inconsiderable energies to the emerging Canada Company. The all-important task at this stage was to obtain the charter before the commissioners sailed for Canada.

The capital of the company had been set at £1,000,000 (10,000 shares valued at £100 each, of which £5 per share had already been paid in). However, until the charter was in hand, the directors could act only as trustees and as such, they

Simon McGillivray (circa 1824)
was a director of the Hudson's
Bay Company and a founding
director of the Canada Company.
As chairman of the Committee
of Correspondence, he devoted
considerable energy to the
company. A man of independent
wealth, he resigned in 1829 over
a remuneration dispute.
Courtesy of LAC C176.

could not apply any of the funds on hand to the work of the company. On the other hand, if the charter were granted immediately, the commissioners would be in a position to pay the first annual instalment to the colonial government – and of more importance to the company, they would be able to commence operations.[56]

The British government was not going to be pushed. The charter would come in due course if the conditions were right.

With the decree from Bathurst, five commissioners, all of whom where London-based at the time, were appointed. The British government chose two, one of which was Lieutenant Colonel Francis Cockburn[57] as head. A very competent soldier, he had been appointed deputy-quartermaster for Upper and Lower Canada in 1818 and, in 1815, had been responsible for the settling and provisioning of the first groups of immigrants and disbanded soldiers to arrive in Upper Canada under Colonial Secretary Bathurst's plan for assisted emigration, of which he was a strong supporter. The second was Englishman Sir John Harvey,[58] a decorated army officer born of modest means. The son of a Church of England clergyman, Harvey went on to be a very competent colonial administrator. The company's Provisional Committee chose Simon McGillivray

and John Galt, while John Davidson was appointed jointly by the British government and the Provisional Committee as the impartial fifth commissioner.[59]
Well-connected in government and business circles, Davidson had extensive
timber trade experience in Upper and Lower Canada through the Quebec-
based firm of Hamilton and Low, one of the three largest operations of its kind
at the time in the Canadas.

Detailed instructions were given to the five commissioners, while McGillivray
and Galt received additional instructions from the company's Court (Board) of
Directors.[60] This would enable the two company appointees to lay the ground-
work for the company in the colony.

The five commissioners began their work in Upper Canada on March 16,
1825, and finished on the 2nd of May. They examined charts of each township
in the province, reviewed the financial returns of each district, and noted lands
repossessed by the sheriffs. They questioned members of both the Executive
Council and the Assembly "in detail." They obtained evidence from "every
respectable and intelligible person known,"[61] including the county registrars.
Their final verdict was 3/6 per acre, currency of Upper Canada, or 3/2 sterling
for the 1,384,413 acres of Crown Reserves and 829,430 acres of Clergy Reserves.
This proposal was rejected, in part because the Clergy Reserves Corporation
would have none of it.[62]

The issue of the Clergy Reserves had been simmering for some time. When
the commissioners' report was made public, the Clergy Reserves Corporation
had a perfect excuse for remonstrating. They stated quite clearly that the price
offered was too low. The government in Upper Canada and the British
Colonial Office agreed – and another inquiry was launched by the secretary of
state for colonial affairs. The official reason offered was "non-fulfilment of the
Instructions either in their letter or in their spirit."[63] Bathurst considered that
the commissioners had made an "imperfect award" and hence, would have to
resume their work to produce one that would satisfy him.[64]

Lieutenant Colonel Cockburn was dismayed and took non-acceptance of
the report as a personal insult. Galt and Davidson were equally upset and a letter
to this effect was dispatched to Lord Bathurst. Wilmot-Horton responded:

> ... Lord Bathurst is not in any degree influenced by the alleged
> inadequacy of the price which they have awarded. He desires to
> be understood as not expressing any opinion whatever upon that

subject. Difficulties of a completely different nature have compelled him to adopt the decision which he had formed ....[65]

Lord Bathurst did not want to admit that price was the issue, nor did he
want to acknowledge the significant pressure on him by the Clergy Reserves
Corporation. That said, neither did he wish to impute any wilful breach or nor
neglect of duties by the commissioners. Accordingly, his official reason for
rejecting the report was that the commissioners had not abided by their instructions. For instance, he noted that they had reported one large general average
for the price of the land, instead of a price for each district. Furthermore, he
noted that their journal had not been properly kept: "the journal ... is ... not
such a journal as your instructions required you to keep, and the deficiencies ...
relate ... to subjects of real and essential importance."[66] He also complained
that the journal did not contain "a minute of *all* deliberations, of *all* occur-
rences and of *all* resolutions." Despite examination of documents for four days,
it was not stated what the documents were, nor for what purpose they were
used, nor proof of what facts they cited:

> Nothing more is to be learnt from them than that four days
> were devoted to reading official papers – a fact perfectly insignif-
> icant without further explanation.[67]

Furthermore, there had been no comment as to how the price was arrived
at – "The result alone is nakedly stated."[68] The registrars of the counties were
not above suspicion because of the frequent practice of inflating the actual price
paid in deeds of conveyance. In fact, the commissioners were specifically criticized for not obtaining the actual price paid for lands sold in ninety-three
townships.

In the final analysis, the commissioners had tried to do too much too
quickly. In the meantime, the shareholders had, by now, paid in a total of
£100,000 – and had not received their July dividend as promised by the
Provisional Committee. Furthermore, the company had still not been incorporated. The shareholders were becoming anxious. The directors, on the other
hand, could not publicly state the cause of the delay which was being imputed
to them. In the meantime, the shareholders were accusing the directors of
neglect. Applications were being received from persons wishing to emigrate, yet

no definitive answers could be provided to them as to when they should leave for Canada to take up the land they wished to purchase from the company.[69]

Further complications had also arisen. Although legally not a company until incorporation, shares were being traded on the market and were dropping in value. John Galt wrote to Wilmot-Horton about the urgent need for a charter. Wilmot-Horton replied that "such sales, if they have been made are contrary to law."[70] By December, Galt complained:

> it is of no use to consider whether the speculations were or were not legal, wise or foolish, the ruin is going on, and the course is not to animadvert on the sanguine imprudence of Individuals but to arrest the progress of the calamity.[71]

It was then proposed that two referees be appointed, with the power of appointing a third in case of disagreement. They were to frame additional instructions "which with reference to the letter and spirit of the former it may be desirable to supply."[72]

The referees were duly appointed – lawyer Sir Giffen Wilson[73] for the government and John Richardson[74] for the company – but it looked as though the controversy would never end. As a member of the Montreal elite, Richardson would be in an excellent position to represent Canada Company business interests. He was an energetic member of the Executive Council of Lower Canada and a dynamic businessman. Furthermore, he oversaw the management of Edward "Bear" Ellice's huge seigneury "Beauharnois" located west of Montreal – and, of course, Ellice was a member of the Canada Company's Provisional Committee. However, their report would not necessarily be the final step. Wilmot-Horton wrote a letter on the subject to Charles Bosanquet,[75] a shareholder and elected chairman of the Canada Company Court of Directors. Bosanquet tabled it at the first meeting of the Court on July 30, 1824. It read, in part:

> I am directed to inform you that his Lordship will not consider himself precluded by the report of these Referees, from referring the whole transaction to the consideration of the Privy Council, for their opinion and final decision ... should he deem it necessary to do so ....[76]

To make matters worse, Richardson became ill and could not fulfill his role as arbitrator. It would be the presence of John Strachan in London that would quickly bring negotiations to a head. Strachan was in London primarily to lobby for the splitting of the Church of England Diocese of Quebec into two parts, with him appointed as bishop of a newly created diocese in Upper Canada. If the company would allow him to represent the clergy in order "to terminate this unpleasant business," Lord Bathurst would appoint Archdeacon Strachan to meet John Galt and one of the commissioners.[77] Company agreement to this proposition was pretty well assured. The lure was held out that if the parties could reach agreement, "his Lordship will not feel it his duty to impede the granting of the Charter." The only alternative was the appointment of a new umpire.[78]

The directors readily agreed to Bathurst's suggestion and a series of meetings finally settled the Clergy Reserves issue. An exchange was proposed.

In return for the 829,430 acres of Clergy Reserves, the company was offered 1,000,000 acres of land adjacent to Lake Huron, within a sizeable parcel of lands bought by the Crown from the Chippewa Indians on April 26, 1825.[79] Because of its significant frontage on Lake Huron, the land was called the Huron Tract (or Huron Block) and had gone into the government of Upper Canada's land inventory as part of the 2,2000,000 acres purchased from the local Native Peoples along the southern half of the eastern shore of Lake Huron. The purchase price was a guaranteed annuity of £1,100, "to paid annually in goods at the Montreal price, but in ratio to the members of the Chippewa Indians who were inhabitants."[80] In 1825, their numbers were estimated at 440. The treaty had been signed in Amherstberg on the Detroit River, in the presence of twenty Chippewa leaders, six representatives of the Crown, plus army and Indian Affairs officers, and interpreters.[81]

The views of the influential colonist John Rolph,[82] a physician, lawyer, politician – a "reformer," interestingly enough – and former adviser to Colonel Talbot, did much to win the company over to this latest proposal:

> Instead of selling straggling Reserves to casual Purchasers, the whole compact tract may be subjected to a system of Settlement free from the causes which have impeded or rendered most expensive all settlements under the immediate care of the Government .... Every Village and Mill will become a centre

around which will be attracted settlers, swarming as it is termed
from all parts of the world.[83]

Drawing on his own experience as a member of the House of Assembly, and
no friend of the government of the day, Rolph continued:

> ... the less the Company is in any manner connected in the
> operations with the Provincial Government, the more satisfac-
> tory will it prove to the Company and, beyond a doubt to the
> Public.[84]

Rolph went on to suggest that the company itself should be responsible for the
survey of the new tract. It should, however, be compensated for this work by
deducting the cost of doing so from the annual payment to the Imperial government. He even went on to suggest that the company should endeavour to
obtain half as much territory again.

The company, however, had had enough of negotiations and, in May 1826,
the directors resolved unanimously to accept the terms as proposed, which, in
addition to accepting the exchange of land, were as follows:

> 1) One-third part or £48,383 was to be expended on public
> works and improvements within the territory, the remaining
> two-thirds only to be actually paid to the government of Upper
> Canada.
>
> 2) The terms "public works and improvements" were under-
> stood to mean canals, bridges, high roads, churches, wharves,
> schoolhouses and any other works undertaken or calculated for
> the common use and benefit of the people in contradistinction
> to works intended for the use and accommodation of private
> persons. Expenditures under this clause were only to be
> approved in cases where such general interests were to be pro-
> moted by undertaking them.
>
> 3) The plan and estimate of every undertaking was to originate
> with the company and was to be submitted to the governor-in-
> council prior to his consent being given to the undertaking as
> to whether it would be received in lieu of payment. If a differ-

ence of opinion occurred, the question was to be referred to the Secretary of State whose decision would be final.

4) Upon completion of any such undertaking the company was to submit a statement of the cost for approval and if over the estimate, it could still be approved provided the amount spent did not exceed the one-third allotted.

5) The block to be allotted was to form a regular mathematical figure as nearly as was consistent with preserving any well defined natural land marks or boundaries.

6) Land granted to the company could be resumed by the government should it be required for canals, roads, the erection of forts, hospitals, arsenals or any other purpose connected with the defence or security of the province.

7) The British government was to survey the block and put a road through from the Clergy Reserves block in the District of Gore at its expense.

8) The company was to have sixteen years in which to pay for the new territory, beginning July 1st, 1827, the payments to be as follows: In the year commencing July 1st, 1826, and ending July 1st, 1827, £20,000; in the year ending July 1st, 1828, £15,000; in the year ending July 1st, 1829, £15,000; in the year ending July 1st, 1830, £15,000; in the year ending July 1st, 1831, £16,000; in the year ending July 1st, 1832, £17,000; in the year ending July 1st, 1833, £18,000; in the year ending July 1st, 1834, £19,000; in the year ending July 1st, 1835, £20,000; and in each of the seven succeeding years, £20,000 (the above represents the amount actually to be paid and does not include the sums to be spent on public works and improvements in the new district).

9) The company was allowed to lay out more money in a given year if so desired.

10) In the year ending July 1st, 1843, the company was to take up all the remaining lands or terminate the contract and abandon all claim to such lands as had not been taken up at that time.

11) If any lands sold were alleged to be unfit for cultivation, arbitrators were to be appointed and lands totally unfit were to lapse to the crown.[85]

The company directors were enthusiastic about the new arrangement. They reported to the shareholders:

> The obtaining of a grant of such an extensive tract of country
> in one mass was beyond the hopes of the Directors, when the
> original contract was made.[86]

The company was now in a position to accommodate all types of settlers, whether single families, to whom any particular part of the province was preferable, or "those associations which agree to settle together for co-operation and society."[87] Although not wanting to excite unreasonable expectations, they outlined in glowing terms the improvements undertaken, or about to be undertaken, in the colony. By this time, construction of the Welland Canal was underway and the building of the Rideau Canal would begin shortly. A system for the improvement of the main or "high" roads was being proposed, and the legislature had been empowered to grant naturalization to aliens, particularly those settlers who had been residents of the United States but who had emigrated to Upper Canada to obtain cheap land and/or to live under the British flag following the American War of Independence.

Initially, the directors decided to concentrate on land sales only and informed the shareholders accordingly:

> The directors in the view which they now take of the prospects
> of the Company strictly confine themselves to transactions
> which will take place with the land only, and therefore exclude
> for the present the consideration of mercantile privileges which
> will be allowed in the charter.[88]

Having outlined the advantages, but still wanting to reassure the shareholders, the directors made a comparison between their agreement and those of "the great American Land Speculations." Specifically, it was compared to the American company of Messrs. Gorham and Phelps in the Genesee country of New York State. The American company, chartered in 1789, had only three years to pay for its land, while the Canada Company was to have sixteen. Gorham and Phelps had to pay for the surveying and opening of roads – the Canada Company had upwards of £48,000 for public works and improvements. Furthermore, in 1826, the American land was worth a minimum of four dollars an acre (or £1 currency) with a general average of about seven dollars (or £1/15 currency) and, of course, the company had negotiated a price of 3s 6d currency or 3s 2d sterling for the land they had agreed to purchase and so, with a proper system of management, the directors concluded, the results "cannot fail to satisfy every reasonable expectation of the Subscribers, Government and the Colony."[89]

Having finally reached agreement, the shareholders were given the option of either continuing in the venture, or opting out and getting their investment back. Eighty-nine percent thought the risk worth taking, with the balance (or 32 shareholders) signifying they wanted out. This latter group was paid a total of £10,950 which represented the original amount paid in, less their proportion of expenses incurred up to June 14, 1826.[90]

On August 19, 1826, two-and-a-half years after the original suggestion of forming a land company was broached, the Canada Company obtained its charter. The company was now in a position to commence operations in Upper Canada under its first commissioner, John Galt.

Five days later, on August 24, company secretary John Galt reported at a Court of Directors meeting that the company had received its charter of incorporation. He read from the Abstract, noting the powers and privileges conferred. Clause number one declared that "the directors, Mr. Galt, and Mr. Freshfield [Canada Company solicitor James William Freshfield of the firm Messrs. Freshfield and Kaye], together with such other persons as shall become shareholders shall be incorporated by the name of the Canada Company." Other clauses declared, amongst other things, that the company's share capital was one million pounds, in shares of £100 each. It also declared that shareholders were to pay calls when requested by the directors, and that shareholders could be sued if in default, including interest. In any case, after six months of non-payment, shareholders would forfeit their shares which could then be sold at a public sale, with the forfeiture declared at a general meeting of shareholders. Calls for capital were not to exceed £10 per share and could not be less than three months apart.

The Abstract further declared the names of the first governor, deputy governor and sixteen directors plus four auditors and noted that they were to continue in office until "the first Wednesday after the 25th of March, 1829." At that time, one-third of the directors and one of the auditors were to be rotated out of office yearly, but they could be re-elected. The governor and deputy governor were to remain in office until 1831. Rotation off the board was to be decided by drawing lots and shareholders of 25 shares or more were eligible to stand as directors or auditors. Five directors constituted a Court or quorum. Two general courts were to be called per year (in June and December), at which time half-yearly dividends would be declared.

With the holding of this Court of Directors meeting, the company was truly ready to move into high gear, but the next years would not be smooth sailing.

CHAPTER 2

THE GALT ERA

Man can have only one paradise on earth, but Galt aimed at having half a dozen simultaneously. He had so many irons in the fire that men doubted whether he could attend properly to the one in which they are interested.[1]

THE CANADA COMPANY'S first commissioner in Canada, John Galt,[2] was an enigma. A man of many talents, he was a well-travelled, big-picture visionary who was at once restless, energetic, scholarly, absent-minded, combative, disorganized, opinionated and well-connected in high places – and he certainly played a significant role in the history of the settlement of southern Ontario. Born in Irvine, Scotland, in 1779, the son of a ship's captain and an eccentric mother, he was primarily tutored at home and in local schools because of family moves between Irvine and Greenock and his less than robust health. He went to work at age 16, first in the Greenock custom house and then, shortly after, clerking in a local firm. In those early days, he formed a literary debating society with two school friends and, through that activity, was introduced to leading literary figures of the day. While working for others during this period, he was able to write quite prodigiously. At age 26, he went into business with a Scottish partner in London and two years later, in 1807, wrote "Statistical account of Upper Canada," which was published in Britain's *Philosophical Magazine*. A business failure in the third year of his business partnership lead to its dissolution at the same time as he was studying law at

John Galt was in his late 40s and "flying high" during the two-and-a-half years he ran Canada Company affairs in Upper Canada – but he regularly ran afoul of company directors and of Sir Peregrine Maitland and his officials. He seemed to feel at least equal to, or next in importance to, the lieutenant governor. *Courtesy of Archival and Special Collections, McLaughlin Library, University of Guelph, H.B. Timothy Collection.*

Lincoln's Inn in London. He was forced to withdraw from his law studies for financial reasons, with an added complication of ill health at the time. In 1809, he tried other business ventures offshore, including endeavouring to circumvent Napoleon's blockade by establishing a trade route through the Ottoman Empire, and travelled to such places as Greece, Malta, Sardinia and Gibraltar. He befriended the great Lord Byron and assisted Lord Elgin in the crating and shipping of the "Elgin Marbles" from the Parthenon in Athens to London. Galt generally travelled in elite circles and related equally well to both aristocrat and commoner when he chose to do so.

In 1811, John Galt returned to London and two years later married 32-year-old Elizabeth Tilloch, the daughter of the publisher of *Philosophical Magazine*. His writings on travel and his production of school textbooks became his primary source of income and, surprisingly, he used as many as ten pseudonyms during this period.

By the early 1820s, Galt had become a known quantity as a parliamentary lobbyist, given his lobbying for support of a large canal project in Scotland (the Union Canal to link Glasgow and Edinburgh) which was being considered in 1819. Furthermore, because of his "Statistical account of Upper Canada" and his having distant relatives in Upper Canada (including the previously mentioned William Gilkison and his friend, the Honourable William Dickson, who was a member of the Legislative Council of Upper Canada and Gilkison's brother-in-law) he was well placed to be chosen to act on behalf of the claimants who were pressing for financial redress for damages suffered during the War of 1812-14.

Galt spent a great deal of his time on the Loyalist redress issue and promotion of the Canada Company concept while continuing to write, and publish, plays and novels. His promotion and lobbying efforts obviously impressed potential investors in the Canada Company since, once the concept had been agreed to, and the company ultimately established, he went, over a six-year span, from being company lobbyist to company secretary to company superintendent/commissioner[3] in Upper Canada.

With the incorporation of the company, a spirit of optimism prevailed – but this was to be short-lived because soon after the shareholders had voted in favour of carrying on, they became increasingly skittish since neither the directors nor the shareholders really knew what it was that they had invested in; this included the monetary value of the Canadian lands they had agreed to purchase. Given this reality, the next logical step was to send a competent person to Upper Canada on a special "Mission of Inquiry and Investigation." John Galt, now 47 years of age and company secretary, was chosen for this task. He left England in September 1826.

On the surface, Galt performed well. However, from the time he arrived in Upper Canada until his ultimate recall and termination in January 1829, he proved to be the most controversial of all the Canada Company's commissioners in its one hundred and twenty-seven year history.

Much has been written about John Galt. Many writers have taken the view that he was harshly dealt with by the directors. However, as one reviews company correspondence and minutes, Galt's letters, his autobiography and contemporary accounts, this clearly was not so. Galt was his own worst enemy! He was well-meaning, but stubborn and, in the final analysis, through lapses in judgment he brought considerable problems onto himself. He was too set on having his own way and forgot, or chose to ignore, the fact that the company could exist only with the support of the shareholders. He also forgot that a little diplomacy goes a long way, especially when dealing with colonial authorities. Perhaps his real problem was an arrogance in relating to colonial authorities, bred from his being so much involved in the start-up of the company that he felt himself on a par with, or failing that, next to the lieutenant governor in importance.

Robina Lizars. Granddaughters of Colborne Township settler Daniel Lizars, Robina and her sister Kathleen wrote *In the Days of the Canada Company* in 1896. It is a wonderful social history of the 1825–50 period of the Canada Company, but it is not entirely neutral in its presentation of information about the company and its employees. *Courtesy of Daphne Davidson & Aileen Davidson Fellowes.*

In order to form an objective opinion of Galt in relation to the company, it is necessary to not only look at his actions through the eyes of the settlers but through the eyes of the Court of Directors in London as well. Although the directors released him for good reason after two-and-a-half years, both Galt's autobiography and Robina and Kathleen Lizars' *In the Days of the Canada*, published in 1896, would indicate otherwise (Helen Lizars, an aunt of the Lizars sisters, was married to John Galt's son, John Galt Jr.). That said, it is apparent that the company did not wish to "take on" Galt following the release of his autobiography. Furthermore, company directors appear to have felt that starting a debate with the Lizars at the turn of the 20th century would not serve any useful purpose for them. In fact, doing so could serve to open old wounds and/or prejudices from people who, rightly or wrongly, may have felt ill-served by the company at a time when company business was going very well. So, the Lizars' publication appears to have gone unnoticed in London and it was only when the company's letter books became available in the early 1960s that the complete other side of the story became available to the public. There is no doubt that the directors were not sufficiently aware of conditions in Canada. On the other hand, Galt, being an independent sort, only attempted to justify his actions after the fact, if queried by the company. He seldom communicated with the head office as to his plans. Furthermore, he was a very incompetent bookkeeper and manager. He seldom asked for instructions. Despite his many talents, he precipitated his own downfall.

Galt's Instructions

Galt's instructions on proceeding to Canada that September of 1826 were extremely comprehensive. Primarily, he was to obtain information as to the best method of disposing of the Clergy and Crown lands (whether by public or private sale, or both), and the terms by which sales ought to be made. Primarily, Galt was to learn as much as he could about the new one million acre territory and was instructed to dispatch experienced persons to examine it. Having taken John Rolph's advice, the directors instructed Galt to make arrangements to have the land surveyed under the direction of the company. The lands, they said, should be laid out according to the topography of the area, with the ordinary government grid pattern avoided (which is at odds with the terms of the agreement of May 26, 1826, which stipulated the use of a grid pattern).

A great deal was expected of Galt. He was to learn as much as he could about the management of the American land companies and decide on the best, and most expedient, method of managing company affairs in Canada,[4] and to take advantage of every opportunity to promote the interests of the company:

> which they [the directors] are convinced is identified with the
> general prosperity of the Colony and will be best promoted by
> a liberal pursuit of these objects of which you are so well
> aware....[5]

Galt's formal instructions set down exactly the type of relationship expected between himself and the company:

> ... he the said John Galt ... shall superintend direct and enforce
> perform and carry into effect ... every such orders directions
> and requisitions as the Court of Directors shall from time to time
> transmit ... in the most ample and beneficial manner. And all
> and whatsoever [he] shall ascertain discover or do in relation to
> the concerns and interests of the said Company ... he ... is to
> report in writing ... without delay to the Court of Directors ...
> and all and whatsoever ... [he] ...shall lawfully do or cause to be
> done under or by virtue of this present Commission or the orders
> directions and requisitions to be from time to time transmitted

or given ... shall be ratified and acknowledged and be held and considered as good and valid by the said company ....[6]

While Galt's directions were a basic plan of action which spelled out the nature of his relationship with the directors, a supplementary article went one step further. It was agreed that although his mission to Canada was for the special purpose of obtaining information, he was to take the initiative of ascertaining which block of land could be got ready for settlers most easily. Having determined this, he was to obtain estimates of the cost of opening a road to and through the block chosen. He was to find a desirable location for a tavern and mill and was to obtain estimates for the cost of their construction. If he thought the work on them should be undertaken immediately, he was given authority to proceed, but the cost was not to exceed £2,000. Finally, he was told that although he had extensive discretionary powers, these powers had to be in line with the directors' overall policy:

> It is unnecessary to add that every step you take will be interesting to the Directors who will hope frequently to hear of your progress and of the result of your enquiries.[7]

The stage had been set for Galt's tenure in Canada.

Policy as Reinterpreted by the Company

At the outset the directors were fully behind Galt. On his appointment as secretary in August 26, 1826, the following was recorded in the minutes:

> ... in voting this appointment the Court [of Directors] are desirous to mark with the strongest feeling of approbation their sense of Mr. Galt's zealous and judicious conduct as Secretary to the Committee for forming the Canada Company, and also as a Commissioner appointed under his Majesty's Commission to value the Lands in Canada.[8]

In order to understand the reasons for the ever widening breach which developed between John Galt and the directors, and hence the shareholders, it is necessary to look first of all at the policy and objectives of the company as developed after the charter had been obtained in August 1826. Galt, ever the idealist, did not seem to comprehend, or wish to comprehend, what certain changes along the way meant in terms of company business procedures in Canada. The company had become realistic while Galt remained idealistic.

The objectives of the company as outlined in the Prospectus of July 1824 were as follows:

> 1. To purchase ... portions of the Crown and Clergy Reserves; to make such other purchases or acquisitions of land as may be found advantageous to the Company; and to work minerals, if deemed expedient so to do;
> 2. To dispose of the Lands, in the discretion of the Company, either to emigrants or to persons previously settled in the country;
> 3. To give immediate employment to emigrants on their arrival in Canada;
> 4. To prepare, by clearing the Lands and by building houses, etc. for the settlement of persons and families to whom the lands are intended to be sold or let, as may be agreed on;
> 5. To make advances of capital in small sums (under the super-intendence at the legal rate of interest in the Colony, which is six per cent) to such settlers on the lands of the Company as may require the same, withholding the titles till the advances shall have been repaid, as well as the price of the Lands;
> 6. To give in this country to persons intending to emigrate information regarding the Lands of the Company, and to facil-itate the transmission of their funds;
> 7. To promote the general improvement of the Colony, whether it be by making inland communications, connected with the lands and interests of the Company, or by extending the cultivation of articles of export, such as Flax, Hemp, Tobacco, etc;[9]

Modified objectives were subsequently developed and reported upon by the company's Committee of Correspondence in April 1827. They represented a general retrenchment from Galt's instructions of 1826.[10] The committee had

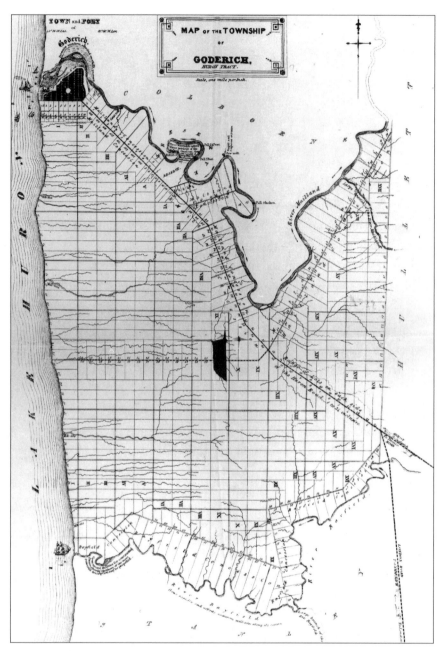

Township of Goderich, circa 1832. Each township was surveyed before the arrival of settlers. The solid black patch denotes a swamp and spring. Alexander Grant purchased it and adjoining land from the Canada Company and, prior to 1854, he built "Spring Creek Mills," a grist and sawmill. The property was purchased by Thomas Trick in 1873.

Courtesy of the Archives of Ontario, F 129 TRMT #238 AO 6488.

examined the subsequent proceedings of the colonial department and the new arrangements which had been "indicated to the Directors" by dint of circumstances.

With the purchase of the lands, the first objective had been dealt with and completed (the minerals question was left open). The second one, to dispose of the lands, would now become the prime objective of the company. Objective number six had required no discussion while number seven, in the words of the committee, was merely declaratory of laudable intentions. It was objectives three (to give employment), four (to clear land and build houses) and five (to make advances of capital) which were now problematic. Yet, the advancement of capital was the only point which was dealt with effectively in committee deliberations. The stipulation of giving employment per objective number three had been evaded neatly by the assertion:

> The Inhabitants of Cities and of manufacturing Districts may
> find employment in the Cities and manufacturing Districts of
> the United States, but it would be acting under a delusion to
> send them as settlers to Upper Canada.[11]

In other words, the company did not want any settlers who were not able to look after themselves on the land. Only those emigrants accustomed to agricultural labour or working as mechanics would be acceptable.

Quite simply, the company viewed its role as being in business to sell land which in turn, from the directors' perspective, was complementary to what the government and people of the province expected, namely, the encouragement of emigration from Britain to Upper Canada. That said, the company's Committee of Correspondence did not feel it necessary to discuss their role in the preparation of land for settlers. They were only concerned with promoting emigration to their lands by way of an extensive advertising campaign. Free passage would not be provided. Furthermore, pauper emigrants would only be considered if the British government paid their passage to Canada *and* on to their intended destination there.

The sole original objective which caused the committee to wrestle with its conscience pertained to the advancement of capital to settlers. It was realized that without an advance in the form of food, seed corn, implements, and perhaps cattle, many settlers would be helpless – and the committee recognized that it could not explain away or evade a condition which had been so explicitly specified. Accordingly, it was agreed to make advances to pauper emigrants, but not in cash. The committee view was that cash advances were neither intended nor admissible. Articles only would be advanced, with "payment" either in labour or in produce.[12] Furthermore, this arrangement would pertain only to the so-called Huron Tract lands. Those on the scattered reserves elsewhere in Upper Canada would have to fend for themselves.

The objectives of the company as reinterpreted in 1827 were quite different in scope from those of 1824. Galt, the idealist and humanist, wished to be guided by the earlier objectives despite the 1827 directive from London. The rift was beginning. Galt had been given new guidelines and the directors expected them to be followed.

Galt in Upper Canada

John Galt landed in New York in November 1826 with his newly appointed "Warden of the Forests," William "Tiger" Dunlop.[13] Their specific initial tasks upon arriving in Upper Canada would be to inspect the lands the company had purchased, determining which lots could be sold quickly and dealing with squatters and timber poachers on company land. As it was also suggested by the directors that Galt meet with officials from counterpart land companies in the United States on his way to Upper Canada, he firstly visited the offices of the Holland Land Company at Batavia and the Pulteney Company at Geneva, both of which were located in western New York State. In the meantime, Dunlop proceeded directly to Upper Canada to begin inspection of company lands in the townships there according to the particular task as outlined by the directors:

> to procure and transmit to the Secretary of the said Company the best possible information of the value state and circumstances of the Company's Lands and particularly the condition and quality of the Timber on those parts of the Company's Lands where plunder has been committed and of the extent of settlements and cultivation in the vicinity of the Company's Lands and also such other particulars as affect the value of the said Company's Lands and are material to the protection and beneficial management thereof.[14]

Dr. William Tiger Dunlop was as medical doctor and the Canada Company's Warden of the Forests. An eccentric, intrepid pioneer, he made his mark not only on behalf of the company, but also as a politician (he won the disputed election of 1841). He wrote a pamphlet, *Defence of the Canada Company*, in 1836 and then turned against the company early in 1838. He died unmarried in 1848 and is buried at Goderich. *The portrait is courtesy of the artist Claudia Elliott.*

Late in December, Galt and Dunlop arrived in York from their separate travels[15] and Galt's controversial two-and-a-half years in the colony had now begun. His time with the management of the two American land companies, which had been operating successfully as large settlement enterprises for several decades, helped him to formulate his plans for the running of the Canada Company operations in Upper Canada. For his part, Dunlop had been able to get an initial sense of the reserve lands that the company would be acquiring.

Tiger Dunlop was as indefatigable as Galt and the two of them were certainly kindred spirits during the period of Galt's tenure as commissioner. Author, army officer and surgeon, Dunlop was born in Scotland in 1792 and had trained as a doctor in Glasgow and London. He arrived in Canada in 1813 in time to treat the wounded from the War of 1812-14, and subsequently had worked on the opening of military roads in Upper Canada. He then retired from military life on half pay and went off to India where he earned his nickname, "Tiger," in an unsuccessful attempt to clear tigers from a tourist destination on an island in the Bay of Bengal. He returned to Scotland in 1820, having become ill, and did some writing and lecturing in Edinburgh and London. In 1824 he met John Galt, having submitted a plan to him to assist potential emigrants. As Galt liked him and the directors were impressed with his credentials,[16] Dunlop joined the Canada Company payroll at £500 per annum, a salary equivalent to the pay and allowances of an infantry captain serving with

Guelph in 1831; the log structure to the right is the Priory. John Galt founded Guelph on
April 23, 1827, in the so-called Halton Block of 42,000 acres northwest of Hamilton.
The land had been allocated to the company because there were no Clergy
and Crown reserves in the Niagara frontier region. *Courtesy of Archival and Special
Collections, McLaughlin Library, University of Guelph.*

the reserves in Canada. He was given the title of Warden of the Company's
Woods and Forests[17] by the directors, but was more commonly referred to as
the "Warden of the Forests."

Upon his arrival in Upper Canada, Galt began to attend to his specific
instructions, which included determining the most suitable location for the
company's principal office. The directors felt that York, as the capital, would be
the logical choice because of the frequent communication which would be
required with the government, but the final decision was left up to him. After
six months in that town, Galt made his choice – Guelph, the community he
founded and named in the 42,000-acre so-called Halton Block, northwest of
Hamilton. While this action is at first puzzling, the reason for his doing so
becomes obvious as the relationship between him and the "establishment" at
York is examined.

An incident which served to haunt him had occurred in the spring of 1825 when he was one of five commissioners from England evaluating the land which would ultimately be purchased from the government. Sometime after arriving in York that spring, he had received an unsolicited a file of newspapers, the *Colonial Advocate,* from the radical reformer and fiery journalist William Lyon Mackenzie.[18] An ardent foe of the Family Compact, Mackenzie was known for using this tactic in an effort to gain converts to his cause, and to increase his paper's circulation. Galt had become his latest target. Galt, unfortunately, was either too new or too naive to understand the colonial politics of the day. He was taken in by Mackenzie's musings, and some time after receiving the newspapers, he sent a cordial note of thanks to the publisher. He then followed up that letter with a second in which one could surmise he may have commented on Mackenzie's grievances including government land policies that were leaving large tracts of Clergy and Crown reserves vacant and the Assembly being ignored by the legislative and executive councillors. Most likely, nothing would have come of the letters had not Mackenzie's office been raided in June 1826 by a group of Tory youth.[19] Newspaper type was scattered and the printing press was thrown into the harbour. In a suit for damages against the invaders, Mackenzie produced the two letters from Galt. This fact was widely reported in the press of the day and, unwittingly, Galt was caught in the middle. It did not matter that he had protested about the use of the exchange of letters as evidence of his connivance with that radical publisher. To senior colonial officials, he was now suspect. Galt was certainly off to a bad start. Lieutenant Governor Sir Peregrine Maitland[20] was not amused.

Maitland would not have to wait long before he would have further cause to be annoyed and frustrated with Galt. This time, it had to do with a three-to-four acre site on Burlington Bay. Galt had picked the site as a very suitable location for a warehouse and supply centre for the Guelph settlement. It was a sensible idea in itself. However, he decided that the site should be *given* to the company because it would benefit all settlers, not just those on Canada Company lands. In making his request in writing to Major George Hillier, the lieutenant governor's influential secretary, he made reference to certain opposition to the company which was apparently coming from "influential persons" in the province:

> … I do assure you that various circumstances have made many connected with the Company not all satisfied with the opposition

Sir Peregrine Maitland was lieutenant governor of Upper Canada from 1818 to 1828. Appointed by Lord Bathurst, he did not like the Canada Company in general and John Galt in particular. While he had the interests of the colony at heart, he was authoritarian and had a hierarchical view of society. He was not popular with settlers. *From D.B. Read,* Lieutenant-Governors of Upper Canada and Ontario. *Toronto: Wm. Briggs, 1900.*

which it is conceived has been shewn towards the general interests of the incorporation, as it now is, from influential persons in this province.[21]

Galt went on to say that he had not personally been of that opinion since the settlement of the Clergy Reserves question. He warned, however, that the directors had political power in England, and, therefore, the colonial government should not act unfavourably towards the company. He concluded:

> … you may tell his Excellency, that among other motives, I was desirous of practically contradicting falsehoods at variance with the uniform tenour of my whole life-falsehoods, the invention of which only served to prove the ignorance of the inventors as to the character of an individual, who from his very boyhood has neither been obscure nor his sentiments equivocal.[22]

He obtained the grant of land on Burlington Bay for the company, but his remarks were little appreciated by either Hillier or the lieutenant governor. This would not be the last time he would annoy colonial authorities.

Galt had not told the company directors about the letter and it was not until the fall of that year (1827) that they became aware of a growing antagonism between their superintendent and the local officials.[23] Obviously Hillier and/or Maitland must have advised the secretary of state for the colonial department William Huskisson, about the contents of the letter, which in turn was forwarded to the directors. Galt was censured severely at a meeting of the company Court of Directors:

> ... the Court disapproves of the tone as well as the substance of
> these Letters, they being alike unauthorized by any proceedings
> of this Court, and the Directors disclaim the opinions ascribed
> by Mr. Galt to "many connected with the Company."[24]

A letter from the directors communicating this displeasure was then sent to Galt, cautioning him as to his future deportment towards the government of the province:

> ... you will carefully avoid the introduction of any allusions or
> supposed opinions or feelings either of any member of that
> Government or the Court of Directors, and the Concurrence
> of the Court in the wish expressed by desire of the Lieutenant-
> Governor that in future all communication upon the business
> of the Company may be strictly official.[25]

In the meantime, the directors sent a letter of apology to Secretary of State Huskisson in an effort to patch things up. It requested the secretary's good offices to smooth things out with Maitland on behalf of their commissioner "in the conviction that Mr. Galt himself will evince anxiety to conciliate his Excellency's favourable opinion and to obtain his confidence."[26]

The next "run-in" with Hillier/Maitland was of a different nature. In the summer of 1827, Galt found himself caught in a "no-win" situation. One hundred and thirty-five destitute emigrants had arrived at Burlington from La Guayra (Caracas) in South America. They had originally been sent there from Scotland by the Columbian Agricultural Association of London, England, but had found conditions intolerable and had left. Through the influence of English officials, they had reached New York. Although originally they were bound for

Nova Scotia, British vice-consul (and Canada Company agent) in New York, James C. Buchanan, steered them to Upper Canada and John Galt. Galt could not very well turn them away, especially when considering that the group included fifty-eight children under thirteen years of age and many women. Thinking they would be a charge on the colonial government, Galt sent them on to Guelph where he arranged to sell them land and build housing. Furthermore, he waived the first payment and employed the men on road building to help them become established. To defray expenses, he then withheld £1,000 from the next installment payable to the colonial government and so notified the directors. The directors were at first sympathetic. In a letter to the secretary of state, Simon McGillivray on behalf of the directors stated that he hoped their superintendent's action would be sanctioned by the government.[27] Huskisson was not sympathetic. Through his parliamentary undersecretary, Lord Stanley, he expressed his "decided disapprobation" at this action and asked for the balance due of £1,000. He also requested the directors "to caution Mr. Galt against adopting measures of his own responsibility which cannot fail to prove serious inconvenience to His Majesty's Government and to the Company."[28] In face of the government's dissatisfaction with Galt's handling of the immigrants and, upon discovering that less than cordial relations existed between Galt and the colonial authorities which they felt had to be patched up, the company backed down and would not support their commissioner. They admitted to the irregularity of the proceedings, apologized for Galt's behavior and directed him to pay the amount due to the government forthwith.

From the government's perspective, a letter from Sir Peregrine Maitland to Lord Bathurst in October 1827, had clinched the matter in favour of forcing the company to pay for the emigrants. Maitland disliked the Canada Company and Galt. He now had a chance to get back at both. He told Lord Bathurst that he felt the company had obtained the land far too cheaply – "these lands were ceded at a price certainly less than one-half than that which I have ever thought …." If the company was allowed to reduce its payments for any reason, he opined, there would be no certainty in the contract made to acquire the land in the first place. From the government's perspective this was serious because of their reliance on the payments to cover fixed expenses as spelled out in the final terms of the agreement between the company and the government. Such a deduction from the amount due could be viewed as a precedent. Maitland then dealt with Buchanan. As far as he was concerned, Vice-Consul Buchanan had acted as an agent of the Canada Company. The immigrants, he pointed out, were meant for Nova Scotia, but had been redirected to Upper Canada. Now, he noted, the company proposed to settle them at public expense on Canada Company land so that they could purchase company property at "ten shillings an acre" – "being a price more than triple given by the company to Government for the same Lands."[29] This was neither reasonable nor just, he concluded.

The whole issue had revolved around Buchanan. Had he been acting as British vice-consul or Canada Company agent? The government had won and unwittingly, Galt was caught – again, but he had certainly not helped himself in the process.

Unfortunately for the company, Galt had issued an ultimatum to the lieutenant governor at the height of the controversy because of a petition addressed to Lord Bathurst by four of the La Guayra settlers. In the petition, the settlers had complained bitterly at being charged "eight shillings" an acre for their land, and between £24 and £30 for the erection of temporary housing. They wanted a grant of land in order to be free of the company.[30] Galt was incensed by the petition and wrote to Hillier:

> C. Coy Office
> 26 Dec 1827
> Sir:
>
> I have to request that you will immediately lay the enclosed [petition] before the Lieutenant-Governor .... It is so strangely at variance with what I have always understood even from the Emigrants themselves that I cannot but consider it belonging to the Singular Series of Coincidences which from the moment I first had the misfortune to Set my foot in this Province has embittered my life.
>
> I beseech his Excellency, in this most strange affair to do justice to my motive and to be in the strongest terms and with the warmest feelings of respect assured that I feel myself compelled to require before 12 o'clock tomorrow as I must then write to the Directors.
>
> First The determination of Government with respect to these misguided people?

Lord Goderich (The Right Honourable Frederick John Robinson). Over the span his 41-year political career, Lord Goderich held almost every important position in the British government, including prime minister. He supported the concept of the Canada Company and, as chancellor of the exchequer, spoke on the company's behalf when legislation was introduced in the House of Commons to obtain its charter. The Town of Goderich is named after him. *Courtesy of the Goderich Public Library.*

Second Whether I am to consider them Debtors to me individually?

And Third that an enquiry be instituted to ascertain on what real grounds they have made this complaint.

You will assure his Excellency that only motives of humanity which even crime can command will prevent me after 12 o'clock tomorrow from giving orders to turn these absurd persons adrift in the Woods ....[31]

Hillier replied:

The Lieutenant-Governor ... commanded me to observe that there are expressions and allusions in your letters which whatever may be meant by them, His Excellency holds it to be equally unnecessary and improper in a Correspondence with this Government and such as if repeated in any further Communication, His Excellency desires you will hold yourself appraised [and] will prevent it being deemed entitled to attention.[32]

Galt apologized,[33] but to Lieutenant Governor Maitland it was one more insult.

Plan of the Town of Goderich. Having been taken to task by Canada Company directors
for calling the first town founded by the company "Guelph" instead of "Goderich,"
John Galt quickly founded Goderich two months later on June 29, 1827. The town plan
was unveiled in 1829, with the Canada Company allocating £3,000
to develop the site. *Courtesy of the author.*

That Galt liked to take matters into his own hands was becoming all too
clear. The naming of the new village in the Halton Block tract of land was a
case in point. With all due ceremony on April 23, 1827, he had founded the
new community.[34] However, without reference to the directors, he called it
"Guelph" in honour of the Royal Family, an anglicization of the German
"Guelf," the surname of the Georges (at the time King George IV the reigning
monarch). Upon learning of this, the directors told Galt that it would have been
much more appropriate to bestow any name pertaining to the Royal Family

upon a more extensive territory.[35] They cited, for example, the new district to be formed out of the one million acre tract (the Huron Tract lands). Furthermore, they had wanted the first community founded by the company to be called Goderich, after Lord (Viscount) Goderich (Frederick John Robinson)[36] who, while chancellor of the exchequer, had been most helpful in getting the legislation to create the Canada Company through Parliament. The directors owed him a debt of gratitude which they wanted to repay. Galt responded that any name change would require an act of the colonial legislature because a number of lots had already been sold, and deeds issued. The directors were not pleased, but were appeased when Galt advised that the next community to be founded would be called "Goderich."[37] Until the matter was finally straightened out, Guelph was referred to by the directors as "the town which you call Guelph and which the Court calls Goderich."[38]

This was not to be the last issue which would mar the relationship between the directors and Galt on the one hand, and Galt and the lieutenant governor on the other. The next occurrence was the celebration at Guelph to mark King George IV's birthday and the first anniversary of the founding of the company. Galt had decided to mark the date (August 12, 1827) with a general holiday and a public dinner. While the celebrations went off as planned, the problem was that Galt, in proposing toasts, had neglected to recognize the lieutenant governor.

Upper Canada at this point had a population of some 100,000 people. The "elite" of the colony were a select few, and many had attended John Strachan's Home and District Grammar School in Cornwall. Most knew one another to a greater or lesser degree and many corresponded with each other on issues of the day. Robert Stanton,[39] for example, was a well-connected businessman, a staunch Tory, public servant and army officer who had been appointed "King's Printer" by Sir Peregrine Maitland in 1826. A Strachan protegé, having attended his school. He had been watching the relationship between Galt and the lieutenant governor with interest. John Macauley[40] was also a well-connected Tory businessman, newspaperman, justice of the peace, militia officer and politician. He too attended the Grammar School and was another Strachan protegé. Stanton and Macaulay corresponded. In writing about Galt, Stanton noted:

> If you see the Que. Gazette, you will find a speech or rather the purport of some observations made by Mr. Galt, at the great fete at Guelph. He seems to have chosen the present opportunity to

Robert Stanton was one of John Strachan's many protegés. He was King's Printer in 1827 when he opined of John Galt: "I shall be much mistaken if he does not turn out to be a very troublesome fellow." *Courtesy of the Toronto Public Library (TRL), J. Ross Robertson, Collection, T16679.*

make some very invidious remarks – I shall be much mistaken if he does not turn out to be a very troublesome fellow.[41]

His words were prophetic!

By late November, however, it seemed as though the tribulations of the summer were being forgotten. Despite Galt's undiplomatic letters and speeches, Maitland began to take a more conciliatory attitude towards him as evidenced by the fact that the lieutenant governor sounded him out for an appointment as head of a proposed regiment (the Guelph and Eramosa Militia),[42] but enthusiasm got the better of the Canada Company commissioner, and he "put his foot in his mouth" once again. In the postscript of a letter written to the directors requesting an cashier/accountant to help him, he reported that he had been *appointed* a colonel in the proposed militia.[43] He then added that he was to have the power of appointing his own officers and observed that such an appointment should set at rest all questions respecting his political conduct.

Galt's optimism after his first year as Canada Company commissioner came through in a letter to a friend in Ireland, Samuel Omay:

> … It would seem that the Directors approve of my endeavour, for they have appointed me as sole manager and raised my

> salary to Two thousand pounds with the hope of increase as
> business expands ....[44]

He added that at last he felt he had achieved his place in the colony. All he asked for now was the power to execute his duties and the privilege to perform them "til I have enabled the Company to pay a dividend." Galt wondered, though, if it was all too good to last. He reported that he was now as well off as any one in the province. He had since moved to a "neat brick house in a pleasant country with a good road" in Burlington, at the head of Lake Ontario, and had three horses, a piggery, a cow, hens and cocks, peacocks and "a fat man cook, a scotch serving lass, a hewer of wood and a gardener."[45] Furthermore, he calculated that he was paid £500 more than the chief justice of the province[46] and that he had "whisky that would obtain renown in Glenlivet." He concluded "all literature is at a standstill, but before the snow comes, I hope to have some little leisure."[47]

To celebrate his first year in Upper Canada and his "appointment" as colonel, Galt decided to hold the "Canada Company's Fancy Dress Ball" at York on New Year's Eve, 1827. It was a wonderful occasion to celebrate, but with an unexpected result. By choosing Lady Mary Willis[48] to be his hostess, he could not have picked a more fitting way to once again insult the lieutenant governor. Galt's wife and family had not yet arrived in Upper Canada as Elizabeth and their three boys – John Jr., Thomas and Alexander (ages 13, 12 and 10 respectively) – were still living in Scotland where the boys were attending Reading School near Edinburgh.

Earlier that year, in April, Lord Bathurst had appointed John Walpole Willis as a junior puisne judge (judge of the Superior Court) of Upper Canada. Willis, who arrived in the colony in September 1826, had very quickly made himself decidedly unpopular with the Family Compact. Haughty and imperious, he clashed head-on with Attorney General J.B. Robinson on matters relating to legal reform (and, in fact, had applied for the position of chief justice of Upper Canada within a few weeks of arriving – a position for which Robinson was also applying). He also counted as his friends such reformers of the day as William Warren Baldwin[49] (doctor, militia officer, justice of the peace, judge, office holder and businessman who was regarded as one of the leaders of the Reform Party); Hugh Christopher Thomson[50] (businessman, printer, office holder, justice of the peace, militia officer, editor of the *Upper Canada Herald* and Liberal representative for Kingston – read "moderate reformer"); John Rolph (physician, lawyer and another leader in the Reform Party); and the reform-minded Americans who had settled in Upper Canada in 1810, Barnabas Bidwell[51] and his son, Marshall Spring Bidwell.[52] Willis also subscribed to Mr. Mackenzie's radical newspaper, the *Colonial Advocate*. Furthermore, Galt regarded Willis and his wife, Lady Mary Isabella Bowes-Lyon (daughter of the 11th Earl of Strathmore),[53] as close friends. Being of noble birth, Lady Mary Isabella would be a rival to Lady Sarah (Lennox) Maitland, the daughter of Charles Lennox, 4th Duke of Richmond and Lennox, and wife of the lieutenant governor. This, in turn, meant that on her arrival in Upper Canada the next spring, the "limelight" would be stolen from "Mrs. Lieutenant Governor" – her social supremacy would be challenged. Because of this, Maitland had actually contemplated removing Willis from the bench as early as December 1827.[54]

Given this scenario, John Galt's choice of hostess for the party could not have been worse.

Word quickly got back to the Canada Company directors about this latest *faux pas*. Again, Galt was reprimanded by them and, shortly thereafter, he submitted his resignation to Charles Bosanquet, chairman of the Court of Directors. In submitting it directly to the chairman, however, he had confronted Bosanquet with the dilemma as to whether or not to present it to his fellow directors. Bosanquet chose not to do so – probably because there was no one else at the time to take over from Galt – and Galt stayed on.

With the snubbing of Maitland through his choice of hostess, Galt heard nothing further about the proposed army commission. In the meantime, however, not knowing what had transpired in York, the company directors relayed to Galt their great pleasure at hearing of his "appointment." It gave a great sense of satisfaction, they said, to witness the appearance of harmony with the provincial government[55] but they would not risk recall of letters to the secretary of state apologizing for Galt's earlier behaviour.

Despite the triangular conflicts, 1827 had been a year of successful development on Canada Company lands. The towns of Guelph and Goderich had been founded and the Huron Tract explored. Surveying of lots had begun. The building of the sleigh (or winter) road from Guelph to Goderich also was underway and company business procedures, such as they were, initiated. The Priory,[56] the residence for John Galt, had been constructed facing the Speed River in Guelph. Galt had engaged a blacksmith and, in order to draw mechanics to the settlement, had built houses for them. Recognizing the importance of educational facilities, he stipulated that half the price paid for building lots was to be appropriated for the endowment and maintenance of schools and he arranged concurrently to build a schoolhouse in Guelph. Houses and "shanties" had also been built to accommodate new settlers temporarily, while storehouses and sheds were erected for company use and a "market house" was underway.[57] Galt had an overall plan for this new community, but it was more than the directors wanted.

For their part, the directors were tidying up loose ends in London. At their July 6, 1827, meeting of the Court of Directors at the London Tavern (their usual place of meeting), the minutes and report of the Committee of Correspondence (chaired by Simon McGillivray) were read, with the various measures approved, including the first resolution which dealt with Galt's salary. The second resolution covered sales of land to employees: "that it is inexpedient to sell or assign any Lot or portion of Land to the Superintendent, or any Agent or Person employed by the Company." Obviously, the directors did not want their people speculating with company land. Other items of business attended to included Galt being authorized to provide financial assistance where requested for the opening of roads in townships where the company had contracted to purchase Crown Reserve lands – but he could only contribute in the ratio of company land to other land and land that was actually occupied. Galt was also authorized to loan funds to the Township of Hamilton for the opening of a road from Lake Ontario to Rice Lake, with repayment, including interest, spread over ten years. With regard to land for churches and burying grounds in the Halton Block, their superintendent was authorized:

> to give possession of a site for a Church or Place of Public Worship and a Burying Ground to any authority in the Church of England, of Scotland, or of Rome, or of any respectable Congregation of any other religious denomination applying for same.[58]

That said, a building had to be completed within limited time period before the site would be granted.

From the directors' perspective, Galt's earlier request for a cashier/accountant was very timely because they were becoming uneasy about the state of company finances in Canada. They now had an excuse to send a person with a solid financial background to discover the exact state of company accounts. The directors specified that the cashier/accountant was to be a co-signatory with Galt on all financial transactions. He was also to be held responsible for implementing and maintaining a proper system of bookkeeping for the company's financial transactions. As well, he was to transmit statements on a monthly basis of all sums drawn "and of the appropriation of same," along with half-yearly returns signifying lands or property sold and all monies received for it – and from whom. In addition, he was to indicate when monies had been received and deposited, and was instructed to break out deposits according to funds received for land purchased versus instalments due and interest paid.[59]

Given Galt's temperament, the directors knew they should choose this person carefully, in that they should send someone capable of working well with him. That said, however, they were irritated with Galt. The accounts were uppermost in their minds; these needed to be sorted out – and fast.

The person the directors picked was Thomas Smith, acting secretary of the company.[60] Besides knowing Galt quite well, Smith had an intimate knowledge of the company, having been with it from the beginning. He knew the kind of financial data required by the Court of Directors on company operations in Canada and, given his background, would be able "to hit the ground running." Smith's role, as outlined by Committee of Correspondence chairman Simon McGillivray, in his recommendation to the Court of Directors, was bound to infuriate Galt – and did. Smith was to be next to the superintendent in authority and emoluments. He was to be permanently resident at the company's principal office – in Guelph – which McGillivray always thought should be located in the capital, York. From Galt's perspective, the choice of Smith could not have been worse.[61]

Emerging Doubts

Aside from problems in Canada, the company was facing difficulties in England. The financial picture was not bright, even though agreement had been reached in 1827 on the lands to be granted to the company in the Huron region (the Huron Tract). While considerable capital was needed on an ongoing basis to enable the company to meet its financial commitments, the company was faced with a decided lack of funds early in 1828. This was primarily because of a pledge given to shareholders in July 1826 which stated that there would be no

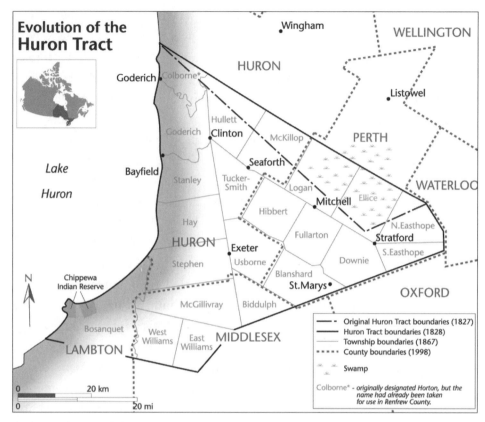

Evolution of the Huron Tract

Legend:
- Original Huron Tract boundaries (1827)
- Huron Tract boundaries (1828)
- Township boundaries (1867)
- County boundaries (1998)
- Swamp

Colborne* - originally designated Horton, but the name had already been taken for use in Renfrew County.

Evolution of the Huron Tract. In 1827, the Canada Company was to acquire one million acres of land in the Huron Tract per an unusual configuration (line with dashes as the northern boundary) in lieu of 829,000 acres of Clergy Reserves. The company's holdings in the Tract were then increased by 100,000 acres in 1828, per the solid line, to compensate for swamps, lakes and land otherwise unfit for cultivation.

Courtesy of the Huron County Museum.

further calls for capital for two years. The company was now faced with insufficient monies to pay interest due in July. A dissolution was discussed.

The government was beginning to doubt the viability of the company, and a call on the shareholders, if made, would undoubtedly prejudice the value of the stock. As it was, the price of shares in February 1828 was nominally £6 on the £10 paid in, or a 40% discount.[62] Would shareholders be willing to risk a further infusion of capital?

If a so-called "license of occupation" (which would indicate the lands in possession of the company in Upper Canada) could be obtained from the colonial government, a call for funds would be justified. In pressing for the license, the directors had a legitimate argument. Under the terms of agreement with the government, all land entered into the company books was to be included in the next quarterly return and then paid for. In the Halton Block, for example, the company had already contracted to dispose of 16,000 acres. Could it be held that possession had been taken of the whole acreage there, and of the Huron Tract as well? If so, a total of £96,525 would have to be paid the next quarter.[63] Furthermore, without the license of occupation, the company had no control over the land not immediately occupied – and squatters and "lumberers" (lumberjacks) were becoming an increasing problem.[64]

In March 1828, the license of occupation issue was solved (the land was granted to the company) and "tag ends" resulting from the two previous agreements were dealt with.[65] An additional 100,000 acres had been granted to the company in lieu of all claims under article twelve of the second contract relating to land unfit for cultivation. The claim for the additional land had resulted from the survey of the Tract, which indicated that in the northeastern part of the territory, an extensive swamp existed in what is now the north half of Ellice and Logan townships. It was believed that this swamp was the principal source of the northern branch of the Thames River. In addition, there were ranges of barren sand hills on the shore of Lake Huron. The additional grant was to take care of those contingencies and any other which might occur[66] by adjusting the upper boundary or proof line from the original proof line. The new line now ran from seven miles north of Goderich through to the northwest point of the Township of Wilmot, a distance of fifty-five miles.[67]

With the important license of occupation solved, and the additional territory granted, the directors could afford to feel a little more at ease. If only the accounts could be straightened out in Upper Canada, the immediate difficulties would then have been resolved.

Smith's Mission to Canada

With Galt's request for a cashier/accountant, and the sending of Thomas Smith to fill that post, the final chapter of John Galt's less-than-stellar relationship with the company directors had begun. Before matters could be straightened out in Canada, Smith was to race back to England ahead of Galt, and both men would be disgraced ultimately. Despite what Galt may have thought about

Smith, he had been sent at Galt's request, and from the director's perspective, to sort out the accounts and help them to understand what was going on financially with their company in Upper Canada. The extent of Smith's detailed instructions annoyed Galt as he felt his authority had been usurped. As far as he was concerned, Smith had been given far too much authority over the accounts – but from the director's perspective, this was exactly what was required. Smith was to be responsible for the accounts. Period! His letter of appointment as cashier and accountant could not have been clearer:

> … the said Company do hereby authorize and empower the said Thomas Smith to call for require and enforce the production of all Letters Books Accounts Papers Vouchers and Documents whatsoever and in whose possession custody or power soever the same may be in Canada aforesaid which in any wise relate to or concern the affairs of the said Company And all and whatsoever the said Thomas Smith shall lawfully do or cause to be done under or by virtue of this Appointment or Commission or any orders or instructions at any time thereafter to be transmitted to the said Thomas Smith by the said Court of Directors as aforesaid shall be ratified and acknowledged and be held and considered good and valid by the said Company to all intents and purposes whatsoever.[68]

From the time the accountant arrived in Canada in August 1828, until his departure in November that same year, the two men were antagonists. That said, it would appear that Smith had had a grudge against Galt and the company from the beginning of his employ in the company in August 1824. He felt ill-used. He had not been on a fixed salary until December 1826, and had always been one step behind Galt. When appointed as company secretary to replace Galt when Galt went on his first mission to Canada in 1825, it was an acting appointment only for Smith, with no additional remuneration. When Galt returned to Canada in 1826, Smith once again was *acting* secretary. From December 1826 until July 1828, his salary had been £300 per annum, yet recall that Galt's had been as high as £2,000 (including commission and living allowance). Smith was plainly jealous, while Galt was resentful of both Smith's intrusions into the commissioner's "space" and his always trying to second-guess him. Smith must have been pleased, though, on the eve of his departure to Canada. His very specific instructions vis-à-vis Galt and the accounts would have buoyed his ego – and his annual salary had been increased to £500 sterling, with a further £250 currency per annum, "in lieu of a Residence and in satisfaction of all other claims"[69]

John Galt would have none of Thomas Smith in the day-to-day operations of the company. Yet how could Smith carry out his instructions without Galt's support and sanction? To run the office effectively, it was necessary that both men trust one another and that they work in a spirit of harmony. What did Galt do? He chose to play practical jokes[70] on the accountant and would not take him seriously. The team had been mismatched and would last only four months. But before leaving, Smith would compile a complete financial statement for the directors. Although (as the story finally unfolded), the accounts compiled by him were not entirely accurate, they were the only ones available to the directors at the time. The directors were distressed.

A meeting of the directors on December 18, 1828, records the following:

> Mr. Galt's letter of 9th December [by mistake for November] was read … announcing that as he could "no longer justify to himself the endurance of the tone and authority Mr. Smith had assigned, and that with the manifest and growing necessity of averting an increase in misconceptions, as well as to afford and obtain the multifarious explanations which seem to be requisite for the right systematic management of the Company's interests has induced him to determine on returning home, in order that a proper understanding may be established and instructions issued by which the management may be harmoniously conducted."[71]

The directors discussed the situation concerning Galt and Smith and decided to wait for Galt's arrival in London "or receipt of statements from him in answer to the Letters addressed to him by Mr. Smith on the 6th and 15th of November, 1828."

But by now the directors had clearly had enough and, two weeks later, on January 2, 1829, they decided "that it is expedient to recal Mr. Galt from the management of the Company's concerns in Canada."[72] The directors then directed the company's Committee of Management to recommend an individual to replace him on a temporary basis and "to suggest needful instructions to be given such person" who was to succeed their superintendent in Canada. Although Galt was planning to return to England, per his letter, the directors were now taking no chances and they resolved to instruct Galt to return to England. At this point, they decided to do nothing further until they could speak to him in person.

Three weeks later the Court of Directors met again. Among key issues discussed, as presented by the Committee of Correspondence, were those pertaining to John Galt. They agreed that Galt's salary of £1,000 would come to an end six months after notice of his termination had been served on him or left at his usual place of abode, and that his allowance and commission were to cease after one month under the same terms. Because his former function and salary as secretary had ceased and merged with his appointment and allowances as superintendent, his recall gave him no claim on his former office.[73] As far as the directors were concerned, Galt was finished – period!

The directors then dealt with accountant Smith. They noted that given the present circumstances, they were in no position to form an opinion with regard to his accounts. However, because the Committee of Correspondence had deliberated on the issues, the directors resolved to do no more than to terminate Smith's agreement of the previous May and give him due notice, with his salary being stopped six months after. His allowance of £250 was taken as paid in full.[74]

All this, of course, was going on without Galt's knowledge, but if he felt hard done by in general, a financial report filed by Smith in November clearly showed Galt's lack of business acumen.[75] Under the heading "Accounts for Guelph," there was merely a general list of disbursements totalling £5,685/2/2 – there were no vouchers, receipts or "connecting statements." A variety of articles purchased chiefly from Montreal and New York was shown, but there were no detailed accounts or statements for those either. The account of land sales was equally unsatisfactory and again, no clear financial statement relating to them had been prepared. The directors could only assume, on the basis of previously reported sales in Guelph, that lots purchased by settlers had, in fact, been paid for. No list of settlers had been sent to London designating those who had paid and those who had not. One further item seemed extraordinary to them. Four thousand pounds had been placed at Galt's disposal for the projected improvements in the 42,000 acre Halton Block. Galt had not asked for any further credit, nor had any been given, yet expenditures had greatly exceeded his £4,000 limit. At Guelph alone, disbursements were stated as being in excess of £6,000 and these had not included the expenses of company clerks, "exploring parties" and surveyors, nor the personal and travelling expenses of John Galt and Tiger Dunlop as they travelled company territory opening up new property, engaging and working with surveyors, deciding on what public works should be undertaken and liaising with government officials in York. The directors raised the question: where had Galt found the money to meet all his commitments? They speculated. Had he transferred company funds for his own use? Had he used deposits received from the sale of land? Furthermore, it looked as though a loss of £2,100 could be expected on the Guelph settlement.[76] To meet expenditures, it appeared that John Galt had transferred funds from the company account in the Bank of Montreal to his personal account in the Bank of Upper Canada. Furthermore, it also appeared that deposits from land sales had been paid to his credit. In what is surely an understatement, the directors commented: "All this appears to have been done contrary to the intention of the Court."[77]

Galt's faulty accounting could have been the result of bad management on his part, or of intention. The directors had no way of knowing. They had already had one bad experience. In the spring of 1827, Galt had confused the Bank of Montreal in connection with £4,444/8/11 currency letter of credit.[78] The bank had been unable to ascertain which sums had been drawn for the use of the company and which were for Galt's own account. Instead of checking with Galt, the bank had debited the full amount, with interest, to the company and credited a like amount to Galt. For his part, Galt had neither notified the directors of this transaction nor accounted for it, and it was only in February of 1828 that the matter was finally straightened out.

To add to the director's confusion, Galt had reported land sales in the colony as follows:

> October 1827 – 40,050 acres
> April 1828 – 72,947 acres (exclusive of the Guelph settlement)

Smith reported:

> October 1827 – 16,000 acres settled for
> October 1828 – 41,227 acres disposed of and settled for[79]

How were these figures to be rationalized? While the directors were attempting to sort this out, John Galt and Thomas Smith were hurling charges and countercharges at one another. Galt had written the directors about Smith's incompetence, "not merely from ignorance but from actual incapacity." Galt called him "arrogant" and "impudent" in presuming to interfere with his *modus operandi*. Smith rejoined that Galt did not understand his own complicated system and that:

> [he] has poured a torrent of vulgar abuse upon my impertinent interference and I am afraid, what is more to be deprecated, the Court may be disposed to blame me for not having interfered more promptly and decidedly.[80]

It was little wonder that Smith was given notice and Galt recalled from Canada.

Galt's Defence

While the directors were puzzling over the Canadian accounts, Galt was attempting to justify himself to them in letters. The explanations, he said, were not offered as a defence, but simply as explanation.[81]

The overexpenditure at Guelph by his calculation was £691/16/2. This he tried to explain away by citing article thirty of the first agreement, the "forfeiture clause." This clause stated that the company was required to place one-half of the lands acquired during each year of the agreement in possession of the settlers in proportion of one head of family for every two hundred acres of such lands. A penalty had been agreed upon in the event of failure to comply, i.e., the company would have to spend thirty-five dollars or £8/15[82] in opening, constructing or improving public roads or bridges on the lands purchased. On the basis of the lands sold and settlers placed, as per the agreement, John Galt calculated that he had actually "underspent" £43/10/9. This, of course, was a fallacious argument and fell on deaf ears. If the statements made and prospects held out during 1827, relative to the sale of lands, had been confirmed by results, no question could have arisen about the forfeiture issue. All that land would have been disposed of by the company and occupied by purchasers. As Galt must have considered his statements well-founded, he could not have thought about, or had any reason to think about, the forfeiture clause until he felt cornered. He was only now attempting to make use of it because he knew his accounts had not been accurate or kept current. The directors replied:

> if the expenses so complained of were deliberately and inten-
> tionally incurred with reference to the forfeiture clause, it is
> singular that the explanation was not given sooner.[83]

Galt should have left well enough alone. The directors were now clearly upset and they would let nothing go unanswered. On the matter of accounts and differences with Smith, they observed:

> inasmuch as it was part of his instructions to investigate as far as
> possible the expenditure stated ... and as he was sent out in the
> capacity of Accountant, these accounts should have been sub-
> mitted to and observed upon [and] remodelled by him, before
> being transmitted to England ....[84]

Galt was suspect now. His suggestion to build a ship to transport settlers from Buffalo to Goderich was criticized as "unwise speculation," yet five years later, when suggested by Canada Company commissioner, Thomas Mercer Jones, the company agreed to build the vessel (albeit with unhappy results). The directors criticized Galt for not having kept them informed on the progress of the settlement at Goderich and, more generally, about the operations in the Huron Tract. They accused him of spending money in Guelph with no idea of the ultimate objective.[85]

To Galt's suggestion that it would be politic for the company to send a note of appreciation to Sir Peregrine Maitland,[86] as he had now been replaced as lieutenant governor by Sir John Colborne,[87] the directors responded that it was hardly necessary. They observed that they had never been "backward" in expressing their respect to him "when any occasion arose for giving an opinion on the subject," and that given the present state of the company, there was no cause for offering much thanks in any quarter.[88]

Galt's last *faux pas* in Canada concerned the December 1828 payment of £8,000 due to the colonial government. Because company bankers in England had made an error, there would not be sufficient funds in Canada to meet the

December payment. Galt for his part knew that there were company-owned government debentures totalling £10,000 in the Bank of Upper Canada's branch in Montreal. However, he could no longer access company funds by himself, having lost his sole signing authority on Smith's arrival, and he did not want to involve the accountant in this episode. So, from his perspective, the only sensible thing to do was to dispatch Dr. Dunlop to Montreal to persuade the bank to advance the monies. Dunlop was successful in doing so, but in persuading the bank to release the funds, he had indicated, on Galt's instructions, that the company's charter could well be forfeited if the December payment was not made punctually.

This whole episode was a "tempest in a tea pot" because, in fact, the original agreement between the government and the company had provided for a penalty, but not forfeiture of the charter, if a payment was not made on time.[89] Furthermore, unbeknownst to Dunlop, the directors had actually written to the bank and Galt some time earlier (a fact Galt had obviously forgotten) authorizing the Canada Company commissioner to request the bank to make the payment on the strength of the debenture. It had not been necessary to apply pressure – and so the bankers must have listened with bemusement (or perhaps concern) as the red-haired Warden of the Forests laid out the case for the company. That said, in the eyes of John Galt, Dunlop's dispatch to Montreal had been absolutely necessary. In his *Autobiography,* he exclaimed "the honour of the Company [had been] saved" by Dunlop's intervention.[90] This episode was but another example of Galt's inattention to detail, coupled possibly with his desire to be seen as indispensable to the company.

In early 1829, Galt was becoming desperate and began to "clutch at straws" in an attempt to vindicate himself. He asked the directors to recall that from the beginning, expenditures for improvements and management had been calculated at a cost equal to the price of lands offered for sale. In doing so, he was ignoring correspondence from the directors on the subject of cutting expenses. While quite logically pointing out that concentrating expenditures at one location (i.e., Guelph) would create a very favourable impression of the company, from the directors' perspective, Galt had not followed instructions. By now, Galt sensed that his explanations would not be listened to and he could only add:

> ... if there do not exist that difference in principle between the Superintendent and the Court, which I think grounds have been given to suppose, it should follow that the misconception

on both sides, as to details, has its origin in my not having been
sufficiently explicit as to the motive by which I was actuated,
and the only apology I can make for this is the plain fact of
having to this hour my attention distracted and my time occupied with small cares and anxieties which however insignificant
individually are by their ceaseless craving and importunity in
collective effect really of the greatest consequence in the
variety of my duties.[91]

Galt would not give up and, between December 31, 1829, and January 24,
1830, he wrote eight long letters to the directors in an effort to justify himself.
Their reply summed up all the grievances against him. They also reiterated the
need for vouchers to accompany the accounts and his "singular neglect" in fur-
nishing them. Galt was chastised for his admitted violation of instructions:

> ... it is the determination of the Court so far as possible to hold
> the reins of management in *their own hand*, and to require at the
> hands of all persons in the Company's service implicit obedi-
> ence of all such instructions as shall be found clear explicit and
> easily carried into effect, instead of a continuance of cavils and
> objections, which are easily enough raised when such is the dis-
> position of persons receiving any instructions.[92]

Galt had made reference to his lost signing authority and the "unexplained
purpose" of looking up receipts. The directors responded that they presumed
the right to retain authority for the expenditure of company funds. They commented sarcastically that the irregularity ("to give it no stronger name") of
spending without authority seemed to be "the rule and their stopping the
irregularity the exception."[93]

In one of his letters, Galt enquired as to why he had not heard from the
directors since July 3, 1828. The terse reply was that his response to a lengthy
communiqué from them of August 21 must have escaped his recollection. They
continued:

> For some time subsequent to that date there were no dispatches
> from you to be answered, and the Court had no occasion to

> address you, being in daily expectation of the long expected
> and often promised accounts, as well as the reports contem-
> plated in Mr. Smith's instructions.[94]

Along the way, Galt had tried to appease the directors by suggesting a novel scheme involving Anthony Van Egmond,[95] a Dutch-born entrepreneur with a checkered past (depending on whom one believes). Van Egmond had recently arrived in Upper Canada with his wife and five children, and had purchased 200 acres from the Canada Company in Oxford East Township east of London. The scheme involved a group of emigrants from Bavaria who had left Europe for Peru after the Napoleonic Wars. Having not succeeded in establishing themselves in South America, they left for New York and were directed to Canada, making their way by wagon train to the Waterloo area via Buffalo and Niagara Falls. Under the proposed plan, the company would pay £90 in cash to each family head, and provide each with 150 acres of land. In return, the set-tlers were to do work for the company for eighteen months. The directors were not pleased, and the plan came to naught:

> … if it be really the case as you say that but for the extraordinary
> circumstances of your present situation you would not have hes-
> itated to enter into a conclusive arrangement with them the
> Court are of opinion that any circumstances which prevented
> such an arrangement being entered into, ought to be considered
> advantageous to the Company so far as it can be called advanta-
> geous to escape a most improvident arrangement.[96]

Galt kept leaving himself open for attack. For instance, he requested a surgeon for the labourers on the Huron Road who were suffering from ague. While the directors gladly agreed to send a supply of medicine to help out, they wondered why a surgeon was needed when Dunlop was right there – "Mr. Dunlop is sometimes called a *Doctor* could not he in a case of emergency give a dose of quinine to a Man attacked with ague?"[97]

Galt talked of the company store at Guelph and his new system of payments for goods purchased. Those who were owed money by the company would buy on account. All other sales were to be for cash only. The directors asked in January 1829:

Where do they find the Cash, unless it be previously supplied by the Company? or what becomes of it after being so paid in for Stores? who are those to whom the Company owes Money? or why should the Company owe Money to Settlers, unless by means unsanctioned by the Court and therefore unprovided for. Is it that the settlers are to be a perpetual charge upon the Company? and that Guelph is to be a scene of perpetual expenditure?[98]

In April, they wrote:

If you alone were the Canada Company, with ample funds at your disposal and if the Court of Directors were your Bankers, your proposed system ... might indeed embrace everything required.[99]

The principal office at Guelph had been established by Galt, and he had inferred that the main business of the company had been done there. Yet, when the accounts were unravelled for the period Galt ran the operation, only £95/15/2 had been received at that office for land sold, out of total company receipts of £1,694/0/1-1/2. Director Ellice chided Galt: more than nine-tenths of the business could have been easily transacted at York. The case was quite different as to payments – "if *they* are to be considered the *real* business of the Company, there appears to have been certainly quite enough conducted at Guelph."[100]

As early as April 26, 1827, the directors had forwarded a formal resolution to Galt for his guidance. It had said that it was not appropriate to attempt any improvements or to incur any expense on the detached lots of the Crown Reserves. Yet, according to the statements in the hands of the directors, a considerable amount had been spent:

If it be replied that a different opinion had previously been entertained and held out to Government or to the Public, and if it be asked with whom did that opinion and some other early dreams of the Company's objects and opinions originate, it would perhaps be difficult for you to disprove your having as

much claim to the paternity of them as to the early discovery
of their fallacy.[101]

The directors acknowledged that had they adopted Galt's suggestions, they
would have had to go along with them. But they hadn't. Galt had been given
amended instructions in 1826 and it was expected that they would be followed,
including no expenditures on the detached reserves. The reason for Galt's
ongoing rebuttals was not very apparent to the directors.

With the tide turning against him, Galt turned to his friend, Joseph Fellows,
whom he asked to inspect the Guelph operation, Galt had first met Fellows
during his exploratory trip to Upper Canada on behalf of the company. As sub-
agent to the Pulteney and Johnstone Estates in New York State, Fellows had
been very helpful to Galt in explaining the operations of the American land
company to him. Perhaps a report by him would put the company's commissioner in a more favourable light.

Fellows did inspect the Guelph operation and wrote a detailed report for the
directors which was sent to Galt for onward transmission to London. The directors were not impressed. They replied:

> The Report which you transmit from Mr. Fellows would have
> been entitled to more attention if the Gentleman had been
> applied to by the Directors to make it, and if he had laid before
> him the whole of the question relative to the Company's
> affairs; but at present he appears in the character of a Judge all
> on one side; and his conclusions and recommendations are cer-
> tainly very much in accordance with that character.[102]

The directors regarded the report as merely an "encomium" (document of formal
or high-flown praise) which echoed Galt's own opinions and one for which,
they said, it seemed Galt wished the American sub-agent to be well paid.[103]

Regrettably for Galt, the reference to a generous payment to Fellows was
once again the result of poor communication between Galt and the directors.
On his way to Guelph, Fellows had collected cash installment payments
totalling the equivalent of some £100 from settlers on Pulteney lands. As he
had not wanted to carry that amount of cash back to the United States, Galt
had offered to help Fellows out by giving him a bill for £100 drawn on the

company account on London in exchange for the cash.[104] However, when preparing the accounts for transmission to London, Galt had merely advised that he had drawn the bill on an existing line of credit without any further explanation. When ultimately presented to the company for payment by the bank, the directors refused to accept it as they felt it must have been a fee paid to Fellows for his report – a report they had not authorized.[105]

Galt had done his best for the company, but it was not good enough. His role in the colony, however, should not be underestimated. He had been a key player in the founding of the company and had the challenging task of organizing its business in Upper Canada. However, it is an untenable situation when a super-intendent of a company, on the one hand, and the directors of a company on the other, are at odds. The directors were finally at their wits end and had no choice but to recall Galt. In a letter to his friend and colleague Tiger Dunlop, Galt reveals that he senses something is about to happen to both him and Smith:

> It would seem that although I have endeavoured to act upon as rigid a code as the Articles of War, it is not enough to satisfy our Masters ... it is not enough that the Commander in Chief endeavours to do his best but that he must be responsible for the conduct of all under him .... As I go to-morrow to Goderich to see what has been done there in order that I may be in readiness for whatever shall prove to have been the decision of the Directors as to Smith and me I send you his precious lucubra-tion .... You will send your remarks ... to the Directors.[106]

Galt would have been pleased as he travelled to Goderich on the sleigh road in late February,1829. He had actually not yet received his letter of recall (he would get it in New York as he was boarding ship to return to London) but wanted to visit the Huron Tract before heading back (per his letter of Nov. 9, 1828 which was read at the Court of Director's meeting of December 18, 1828) in an attempt to set things straight from his perspective. The Huron Tract was now receiving settlers and a spirit of optimism prevailed in the new town on Lake Huron. He stayed with Tiger Dunlop, in his log "Castle" overlooking the Maitland River estuary, and likely would have mused about his first trip to the town site by water and his "christening" of Goderich. But he also would have been complaining to Dunlop about his treatment in the hands of the company and would have discussed how Dunlop should support him in a letter which he asked Dunlop to send to the directors. On his way back to Guelph, Galt visited with his friend Anthony Van Egmond at his farm near Seaforth. In Guelph, he bade farewell to the settlers there and then left for York and New York where he sailed for Liverpool on April 5, 1829.

Following Galt's visit to Goderich, Dunlop did write to the directors and naturally defended Galt, but it didn't help – the directors had made up their minds.

The *Montreal Gazette* of March 30, 1829, reported tersely:

> Mr. Galt the author, who has hitherto acted as Manager for the Canada Company has been dismissed from his Office by the Directors. Mr. Galt's character is unimpeachable, but it appears he was too fond of having his own way, and this gave offence.[107]

The directors were determined to retain under their control the authority as well as the responsibility for the Canadian operations. If the initial instructions were not clear to him on this point, subsequent correspondence from the directors had made them abundantly evident.

Perhaps too much had been expected of John Galt. The company report of 1827 tells of a desire to combine economy with energy and activity. In the hope of being relieved of a costly establishment through legislative action, the directors had deferred the appointment of a board of commissioners as contemplated by the charter. The net result was that Galt's duties had increased. He not only had to collect and transmit local information, but also had to manage the company's growing operations in Canada.

Galt was not unaware of his workload. This he did not mind, but he did want it clearly understood, he said, that he was not merely preparing an easier office for another. Early on in his tenure in Upper Canada, he had asked for the director's confidence and to be excused for errors in judgment. This the directors had been willing to grant. His resignation had not been accepted in 1828, in all probability, for this very reason.

Why, then, had he been recalled and dismissed? There were three very good reasons: he could not abide the colonial officials, his bookkeeping was appalling and he did not respect the authority of others, especially his directors.

CHAPTER 3

A CRUCIAL YEAR

Galt lost the confidence of the Company directors in London
[and] ... was unceremoniously replaced.[1]

WITHOUT A DOUBT, the most crucial year for the Canada Company was 1829. The decision to recall John Galt had been made. The shareholders were uneasy and there was a definite possibility that the company would have to cease operations and surrender its charter. Furthermore, three directors would resign that year – Messrs. Hodgson and Williams early in the year, and McGillivray in September.[2] The company's endeavours in Canada had not been overly successful. If the company was to survive, or even if the decision was made to terminate the charter, a new commissioner would have to be appointed and sent to Canada in the meantime.

In considering what to do following Galt's departure, there were three issues to resolve: the naming of his replacement(s), the salary to be offered and the duration of the assignment. In this regard, the directors were guided by Edward "Bear" Ellice, merchant banker, landowner, politician, Hudson's Bay Company director and influential member of the Canada Company's Court of Directors. Born in London in 1783 into very favourable circumstances, Ellice had inherited the family's significant commercial empire in 1805 upon the death of his father. It included significant land holdings in the United States, British North America and the West Indies. Ellice had gone on to become a prominent merchant banker and shipowner, trading in fur, fish, sugar, cotton and general merchandise in North and South America, as well as trading into the East and West Indies and Europe. In 1809, he married into one of Britain's most powerful dynasties (his wife, Lady Hannah Althea Bettesworth, who had been widowed, was the younger daughter of the 1st Earl Grey). In the 1820s, he became a principle spokesman of the Whigs on economic questions, having been elected into the British House of Commons in 1818. For over forty years Ellice was an extremely active member of parliament. He was also a key figure behind the merger of the NWC and the HBC into the reorganized Hudson's Bay Company in 1821. His inherited lands in Canada and New York State at one point totalled 450,000 acres, including the Beauharnois estate west of Montreal, consisting of 324 square miles on the south shore of the St. Lawrence River. His other investment holdings in the Canadas included the Welland Canal Company, the Canada Company and the Bank of Upper Canada, plus government securities and mortgages which he took back on land sold. Ellice was a passionate advocate for the union of Upper and Lower Canada and was universally known by the nickname of "the Bear" because of "his wiliness ... rather than any trace of ferocity."[3]

Given Ellice's significant involvement in managing his personal assets, his other business dealings and his not insignificant political activities, it may seem surprising that he was able to make such a personal commitment to the Canada Company in its early days. In fact, he was only able to do so because he had lost the election of 1824 and had only re-entered the House of Commons in 1830 – in the meantime, he had been rotated off the Canada Company Court of Directors in March 1829. Ellice was actually highly critical of the government's policy that created the company in the first place. He felt a more realistic system of land sales could have speeded up development in Upper Canada. It was because of this slow pace of development that Ellice had been a driving force behind the abortive Union Act of 1822 and had continued to press for the union of the two colonies which finally happened in 1841.

The first item of business on Ellice's agenda following Galt's release from the Canada Company was to recommend to his fellow directors that the former commissioner be replaced with two commissioners – one from England and one from Upper Canada. For the English recruit, he strongly recommended Thomas Mercer Jones.[4] Born in England of Irish heritage in 1795, Jones had worked under Ellice for twelve years in the London-based mercantile firm of

Edward Ellice was a highly successful businessman with considerable inherited wealth and land in North America. He was a director of the Hudson's Bay Company and a founding director and deputy governor of the Canada Company in its early days. He was also an elected politician in England (a Whig) and a spokesman for that party on economic questions. *Courtesy of LAC 2835.*

Ellice, Kinnear, and Company, and the financier/businessman had the highest regard for him:

> … an intelligent, excellent and steady young Man – about thirty years of age – possessed of some patrimony of his own (between £4 and £5,000) having great experience in accounts – being steady to a proverb.[5]

Because of Jones' "integrity and strict obedience to instructions," Ellice was even prepared to act as surety. Given Ellice's stature, and his knowledge of Jones, the directors looked no further and so agreed with the recommendation. He seemed eminently well-qualified and, being unmarried with no other commitments, could proceed to Canada on short notice.

A second commissioner was suggested by Ellice because of the importance, in his view, of having, as one of their men in Canada, a person with "local knowledge and personal influence" in the province. Such "a person of station" could help restore confidence in the company's management, he said.[6] His colleagues concurred and the search for the second commissioner was narrowed down to two finalists.

Both were highly regarded and understood land administration issues. One was the active and well-connected Executive Council member George Markland, while the other was Legislative Council member and Bank of Upper Canada president William Allan. Both were resident in York and, as members of the Legislative Council, were entitled to the designation of "The Honourable." This title was considered important for the prestige of the company. Their qualifications were outlined:

> Mr. Markland: A younger Man, whose time would probably be more at his own disposal, and who had already considerable knowledge of the Company's affairs; having been in England, and in communication with the Colonial Department during the discussions with Government in 1825 and 1826, and having applied to several Directors for employment as Commissioner in the Company's service, at the time of Mr. Galt's being sent to Canada.[7]

Markland was born circa 1790 in Kingston and educated at John Strachan's school in Cornwall. In 1810 he was referred to by a leading cleric of the day, the Reverend John Stuart, as "a good, indeed an excellent young man" who wished to enter the Church of England ministry. John Beverley Robinson, on the other hand, while describing him as "a good fellow and very friendly" said he preferred "a person of his age [to be] rather more manly and not quite so *feminine* either in speech and action." Despite the prejudices of the day, this latter observation did not hold him back from being appointed to increasingly important positions. He served in the Frontenac Militia during the War of 1812–14, ran unsuccessfully for election in 1820 in the riding of Kingston and, a few weeks later, was appointed to the Legislative Council. Two years later, he was made a honourary member of the Executive Council of Upper Canada which, in turn, led to him becoming a regular member of the Council five years later. In 1822, he was also appointed to the provincial board of education and, in 1828, was appointed secretary-receiver of the Upper Canada Clergy Reserves Corporation which administered the leasing of the Clergy Reserves. That same year, he also became registrar of King's College and was involved, the next year, with Sir John Colborne in the creation of Upper Canada College in Toronto. In May 1833, he was appointed to the prestigious position of inspector-general of public accounts.[8]

> Mr. Allan: ... a Man of large Property and well known habits
> of business, resident at the seat of Government, and President
> of the Bank of Upper Canada; whose advice and countenance
> would be of high importance, but who from his own avoca-
> tions could not be expected to devote much time to the con-
> cerns of the ... Company.[9]

Born in Scotland circa 1770, William Allan[10] had come to Canada about 1787
with little formal education. Initially he worked in Montreal in the firm of
"Bear" Ellice's father, Robert Ellice and Company, a trading firm which was
subsequently reorganized as Forsyth, Richardson and Company in 1790. This
experience started Allan on his way to becoming successful businessman. In
1788 or 1789, he was transferred by the company to Niagara-on-the-Lake, an
important trans-shipment point on the route to the upper lakes located at the
juncture of the Niagara River and Lake Ontario. He gained influential friends
along the way while learning the "ins-and-outs" of merchandising and trading.
In 1795, after obtaining a town lot plus two hundred acres in York, he moved
to the new capital of the colony with its less than 700 people and became the
agent for Forsyth, Richardson and Company. In addition to running this
agency, he had also been involved as a merchant in two other businesses. As if
that were not enough, he held down such government positions early on as
license issuer for marriages, shops and taverns, district treasurer of the Home
District of the colony, collector of customs at York, district inspector of flour,
potash and pearl ash, postmaster at York and returning officer for a number of
provincial elections. In 1803, he was treasurer for the building fund of St.
James' Church, York and people's warden there when John Strachan became
rector in 1812. Strachan and Allan were to become close friends.

Allan had also been involved in the military and became a militia officer in
1795. He fought in the War of 1812-14 and became commander-in-chief of the
militia at York. For his efforts, he received 1,000 acres of land. Following the
war, he gradually moved on from being a leading merchant and local office
holder to becoming a prominent financial figure. Although pressed by many to
run for elected office, he always refused. In 1818, he became agent for the Bank
of Montreal in York and, in 1821, following the incorporation of the Bank of
Upper Canada, he headed the subscription committee of the bank. The next
year, Allan gave up his trading interests and became a director of the bank,

William Allan was born circa 1770 at "the Moss," near Huntly, Scotland. When he became a co-commissioner of the Canada Company in 1829, he was already a prominent York-based banker (president of the Bank of Upper Canada), an exceedingly wealthy landowner, a successful businessman and a member of the Executive and Legislative councils of Upper Canada.

Courtesy of Toronto Public Library (TRL), J. Ross Robertson, Collection: JRR 3541.

Moss Park, Toronto. William Allan had little formal education. While a "self-made man" who lived in elegant surroundings with his family in Moss Park (Sherbourne Street south of Shuter Street), he had a sad personal life. Nine of his eleven children died before reaching 20 years of age. Only one son, George William, survived him. He died in 1853.

Courtesy of the Toronto Public Library (TRL), J. Ross Robertson, Collection: T 11099.

along with John Strachan and other notables. That same year, he was elected president of the bank, a position he held until 1835 when he resigned (except for one year when he was travelling). In 1824 Allan became a director of the Welland Canal Company; in 1825 he was appointed to the Legislative Council and, in 1827, he served as a legislative councillor on a government committee dealing with navigation on the St. Lawrence River.[11]

Allan was approached to become a potential Canada Company co-commissioner when he certainly had more than enough to do, but undoubtedly it was his close connection to Ellice that persuaded him to consider it. While Allan had given up his trading interests, he had taken on the responsibility for Ellice's land speculation and financial interests in Canada. Additionally, Allan had property himself in almost every district in the province and so encouraging settlement on the former Crown Reserves now being settled by the Canada Company would serve him well too. Needless to say, the combination of his being so well-connected in Upper Canada and his flair for incisive analysis in matters financial would stand the company in good stead.

From the company perspective, Markland was the preferred candidate, but they were not certain he would accept their offer. Allan was to be their alternate. Because of this uncertainty, Jones took two sets of instructions to Canada – one for him and Markland, and the other for him and Allan which was to be opened "*only* in the event of death, or refusal, or other objection preventing the appointment of Mr. Markland."[12] Markland turned down the company's offer and so Jones pulled out the second set of instructions.

Undoubtedly Markland decided not to take the Canada Company up on its offer because of his not-insignificant commitments at the time. In retrospect, his refusal to do so turned out well for the company because it seems that, as early as 1835, he was having sexual liaisons with young men. When this surfaced publicly in 1838, an inquiry that had been looking into the allegations in the meantime was quietly dropped in return for his resignation as inspector-general. Fellow officers then pushed him into resigning his commission as colonel in the Frontenac Militia. He was not re-appointed a legislative councillor in 1841 (he had already resigned from the Executive Council in 1836). That same year it was then discovered that as treasurer of the school lands, while a member of the board of education, he was in default of £5,000 for the period covering 1831-38. Furthermore, he was nearly caught in a civil suit because of school funds being diverted to build Upper Canada College. On the default

Thomas Mercer Jones was named Canada Company co-commissioner in 1829. Born in England in 1795, of Irish descent, he married Archdeacon Strachan's daughter, Elizabeth Mary, in 1832. He moved from York to Goderich in 1839 with his family, gradually developed "localitis" and was terminated by the company in 1852. He was then agent for the Bank of Montreal in Goderich until his wife died in 1857.

Courtesy of LAC C-098835.

issue, he did not deny responsibility and the government was reimbursed over a period of time. Was the default because of careless accounting practices in the mould of John Galt, or something more serious? In retrospect, the Canada Company was most fortunate that Markland did not accept their offer.

On the matter of salary and the duration of the assignment for the new commissioners, company directors agreed with Ellice's recommendations that the appointments be for one year only, with the salary for each commissioner set at the rate of £400 per annum. Jones was also to receive an additional £250 sterling. This was allocated in lieu of a residence in Upper Canada. It was also to be used to defray the cost of his voyage to and from Canada and for travelling expenses while in Canada. Edward Ellice obviously believed in Jones as he agreed to post a £5,000 surety bond for his protege.[13]

Jones' and Allan's one-year "temporary arrangement" was destined to become permanent. The two co-commissioners were to guide the fortunes of the company in Canada until Allan left the firm in 1840 in his seventieth year (his resignation and the company's letter asking him to retire crossed in the mail), and Jones was released in 1852.

Skeptical Shareholders

With many of the shareholders entertaining doubts about the future of the company, the directors were placed in a difficult position. At a general meeting

Villa of Thomas Mercer Jones in York (Toronto). Following his marriage to
Elizabeth Mary Strachan in 1832, Jones built this villa in York on a lot severed from
Archdeacon Strachan's property. At the corner of Front and York streets, the villa was
designed by John G. Howard. Jones had the reputation of being somewhat of a cad and
spent quite long periods each year working out of Goderich before moving there with his
family in 1839. *Courtesy of the Toronto Public Library (TRL), J. Ross Robertson Collection: JRR 814.*

of the Court of Proprietors (shareholders) on March 31, 1829, they had been
very surprised by the attitude of the shareholders. In January, Galt's recall had
been fully explained as had the rationale for the temporary appointment of the
co-commissioners. The directors had not wanted to deceive themselves nor
mislead the shareholders and could see no reason for the company not continuing. Yet many shareholders were entertaining doubts about the viability of the
company and its operations in Canada. They kept pressing for answers. Was the
land really worth the contracted price? When would be the next call for capital?
While the directors knew a good bargain had been made in the purchase of the
land, they could not say when a call would be issued for additional capital.
There were too many unknowns.

In spite of the directors, two resolutions were tabled, and carried:

> 1. The directors be instructed to enter into a negotiation with
> His Majesty's Government to reconsider or relinquish the
> present agreement.

> 2. That it is the opinion of the present general court that no
> further calls on the proprietors should be made without the
> previous consent of another general court.[14]

The directors were clearly stunned. Not only did the second resolution run counter to the company's charter, it would prevent the possibility of a successful negotiation with the government as contemplated by the first resolution. How could the company make good its payments to the government, or even offer reduced payments as an inducement to grant the desired modifications of the contract?

The real issue for the directors which impacted on the shareholders' view of the company was that they were unable to provide an accurate financial statement – and it was compounded by the fact that the directors were still not fully informed of Galt's various contractual obligations. Thomas Mercer Jones and William Allan had been instructed to prepare a report on the state of affairs in Canada, but until it was received, little could be relayed to the proprietors.

At that March meeting, the directors had reported on the business of the company as honestly as they could from the information on hand. Of the 270,000 acres of Crown Reserves currently in possession of the company, they indicated that only 64,000 acres had been sold. This news was hardly encouraging for shareholders. The company directors could only plead that with better management, sales would improve. They felt the experiment in Canada had not been given a fair chance and, in the present circumstances, were finding it extremely difficult to make a case to the government and parliament to ease their plight.[15] They had already made representations which had not been well-received. Any hope of modifying the contract was pretty slim. It was realized in government circles that the company had made a very good bargain. It would be no loss to the colony if the company gave up its charter, despite government dependence on company payments for operational needs. With land sales increasing, these costs could be met.

Since the shareholders had forced the hand of the directors, a proposal to modify the agreement was presented to the British government. The new agreement would entail the company relinquishing its existing contract for the purchase of the remaining Crown Reserves. However, it would exclude those portions of the townships of Guelph and Wilmot (now in Wellington County) not already in the company's possession or approximately 120,000 acres for

Sir George Murray was Britain's
second-most-decorated soldier after the
Duke of Wellington who, in 1828 while
prime minister, "parachuted" Murray into the office
of colonial secretary. While weak and vacillating in
this latter capacity, Murray called the bluff of the
Canada Company in 1829. He was replaced as
colonial secretary in 1830. *Courtesy of the National
Portrait Gallery, NPG 4319.*

which the Canada Company had paid some £20,000. The balance of the funds already paid (£22,000) would be applied to the purchase of the Huron Tract lands. If this proposal were accepted, it was estimated by the directors that £12/10 per share would be required on top of "a like amount" that had already been paid in by them.[16]

With this proposal, the directors were confident that the company could carry on successfully. Without the scattered Crown Reserves company management expenses would be reduced and the Crown Land Commission would be able to sell lots in these areas to settlers for less than the company could sell them because they already had agents in those areas. Any jealousy between the province's land granting department and company agents would be removed.[17]

The British government would not listen. The charter from the government's perspective had been granted chiefly for the purchase of the Crown Reserves and had been recognized as such by an Act of Parliament. Any deviation would require a surrender of the charter. Secretary of State for the Colonies Sir George Murray[18] mused "how much reliance could be placed on a new contract?" The threat by the company that they would be unable to meet its next installment of £7,500 rang hollow in his mind. The directors had thought as much even if the shareholders were not on side:

> It might easily have been anticipated that the determination at
> *once* to stop all further supplies and thereby to embarrass the

financial arrangmts which Govermt. were known to have made, would be the most effectual means to prevent the success of any application on behalf of the Coy. And as connected with the Second Resolution, the first was really instructing the Directors to accomplish an impossibility.[19]

The dissident shareholders had won the day, and the directors had to follow through with developing the new proposal. To the directors, the proceedings of March 31 had been calculated to destroy the existence of the company as a public body. However, they felt it was their duty to protect the ultimate interests of the shareholders at large. In doing so, they offered to resign if: a) the majority were determined to persist with directive that the directors were to press the government for a renegotiation of the original agreement, or b) a better qualified group emerged to conduct the affairs of the company. Until such time, they would carry on as they saw fit.[20]

The exchange of the Clergy Reserves for the Huron Tract lands was at the heart of the problem for the dissident shareholders, as was the fact that the government was giving away land *gratis* to settlers elsewhere. Company shareholders had not really appreciated the opportunity presented by the acquisition of the Huron Tract lands. Or if they did, they undoubtedly realized that the return on their investment would take longer to materialize than if the company were merely selling former Clergy Reserves scattered throughout the townships in Upper Canada without attendant infrastructural obligations. Furthermore, it would appear that the directors had not counted on competition emanating from the sale of the Clergy Reserves, nor indeed on a government land giveaway program.

Clearly, the directors had not expected shareholder negativity. By the spring of 1829, because of the loss of confidence in the company, the per share value of company stock had sunk to a low of £1/10/0 from a paid-in value of £12/10. Many shareholders had sold out at a loss just to escape the liability of answering future calls for capital.[21]

In a letter to R.W. Hay, undersecretary for the colonies, the company's chairman, Charles Bosanquet, explained the Canada Company's predicament:

While the credit and the prospects of the Company were ... deteriorating in England, they have had to contend in Canada with an unlooked for competition in their sales from the Commission appointed to dispose of the Clergy Reserves, of which they have been deprived, as well as other lands called the School Reserves and besides that, the Provincial Government has continued to make free Grants to Settlers.[22]

Emigration to Canada, he noted, had decreased and many would-be settlers had gone elsewhere, including Australia, because of settlement schemes there. Furthermore, grants of land, complained Bosanquet, had been given to speculators and settlers on more advantageous terms.

On the basis of the accounts as drawn up by accountant Thomas Smith just before he left for England in November of 1828, Bosanquet outlined the company's financial position:

Expenditures:	
Payments to the Crown (to date)	£42,500
Expenditures in Canada –	£37,500
chiefly local improvements	
Total	£80,000
(excluding payments and charges in England)	
Receipts:	
Sales of land	£29,000
Deposits and instalments received (cash)	£ 6,000
Total	£ 35,000[23]

If the company could concentrate on the Huron Tract only, a direction and impetus could be given to settlements "which would be of great public importance." The Guelph settlement was cited as an example of the company's endeavours in this respect and the financial embarrassment in the colony which would result if the company failed was noted.

By December 1829, the directors had placed two propositions before Sir George Murray – either he consent to the half-yearly payments being reduced, or the company be at liberty to surrender its charter.[24] Murray accepted the latter and said appropriate instructions to this effect would be issued to Sir John Colborne, the new lieutenant governor of Upper Canada. The directors' hurried

response was that they could do nothing further until a meeting of the shareholders was held, adding:

> It now ... occurs forcibly to the Directors that they omitted to
> suggest a consideration which had been previously adverted to,
> ... instead of winding up the affairs at once as proposed, the
> Company should go on for two or three years certain under the
> present Contract, ... with the liberty at the expiration of that
> period either to proceed to the end of the Contract, or to close
> the concerns in the manner now proposed.[25]

Sir George Murray had called the bluff of company management and shareholders. He knew something about the Canadas, having spent a three months
there in 1814 (primarily Quebec City and then as provisional lieutenant governor of Upper Canada) and had been following developments with certain
interest from England. Although a weak secretary of state who had been parachuted into his position and would last only two years there, (1828 to 1830), he
was considered to be "a good man of business" and was perfectly aware that if
lands in the United States could be readily sold for the equivalent of 6/3 per
acre there, Canada Company shareholders had no right to complain of their
bargain. Sir John Colborne had backed him up. Murray had not wanted the
company to exist in modified form. The government had the upper hand, and
had won.

CHAPTER 4

A DECADE OF ALLAN AND JONES

His [Allan's] reputation rested on his success in business.[1]

THE NEW CO-COMMISSIONERS, William Allan and Thomas Mercer Jones, had personalities and business habits totally unlike those of John Galt, and they understood and acknowledged what was expected of them. Jones was to devote his full time and attention to company concerns, while Allan was to perform as many of the jointly assigned duties as would be compatible with his other "avocations." Allan was not expected to travel. That would be left to Jones. Between the two of them, however, no distinction was made as to precedence or authority.[2]

With the appointment of Allan and Jones and the subsequent "go-ahead" of the shareholders to continue operations, the company was entering a new phase. The first objective was to be profitable, with adequate returns for the shareholders. This could only work by selling the land purchased at an enhanced price. This fact was made perfectly clear to the commissioners from the start.

Initially, Jones and Allan were given little latitude. It was much simpler to rule from London with a tight rein and then ease off as conditions warranted. It was most important to restore confidence in the company in both England and Canada, to regularize office procedures in the colony and to trim expenses until the accounts could be analyzed.

Daniel Lizars, a friend of John Galt from Scottish days, was a wealthy settler who lived in Colborne Township. A leader in the community and active member of the Colborne Clique, he was critical of the company and enjoyed baiting Thomas Mercer Jones, but went to his defense when Jones was terminated by the company in 1852 over the railway issue. *Courtesy of Daphne Davidson & Aileen Davidson Fellowes.*

Many inhabitants of Upper Canada were critical of the company. It was, of course, very easy to blame a large land company for any colonial ills, and to use it as a means of getting "at" the government. Furthermore, there was speculation in Upper Canada that the directors were intentionally running down the value of the shares on the market to enable their purchase at a significant discount. Company records in Canada were in a mess and settlers were wondering whether they would ever be able to obtain their deeds for land purchased.

It was in this setting that Thomas Mercer Jones and William Allan began their work. No one really knew whether the company would survive the crisis period of 1829, and the commissioners' task was being made all the more difficult by the apparent free spending of John Galt. The commissioners and directors would need to work closely together through the decade of the 1830s. They would also have to deal with settlers' complaints, whether of their making or not, and run the company business from profit-oriented objectives. Jones, particularly, was in a difficult position because he had to bear the brunt of criticism from the so-called Colborne Clique[3] (or Colbornites). These were the residents of the relatively small 35,460 acre Colborne Township just north of Goderich. Many had been friends of Tiger Dunlop and John Galt in Scotland and had been induced to come to Canada by them. They were well-educated and delighted in criticizing the company, their particular gripes centring around bridges, roads and mills. But they were also annoyed at John Galt's dismissal. The group included Henry Hyndman, an Englishman who purchased 800 acres in Colborne Township and arrived there in 1832 – he acted as a

"Lunderston" – Henry Hyndman's home. Henry Hyndman, an Englishman from London, purchased 800 acres in Colborne Township in 1834 and built his home there with an extensive library. He was a humanist. As many as 32 new arrivals at any one time would stay with the Hyndmans until able to build their own accommodation. He died prematurely in 1844 after being thrown from his wagon. Active in local politics, he became sheriff in 1842. *Courtesy of Library and Archives Canada / C – 098835.*

returning officer for the 1835 election, and later became the first sheriff of the Huron District; Daniel Lizars, who arrived in Colborne Township in 1833 from Edinburgh with his mother, wife and seven children[4] – he became acting district clerk in 1842; and John Galt Jr.,[5] who had gone back to England with mother and brothers in the summer of 1829. Galt Jr. returned four years later at age 19 to take up farming and, in 1840, he married Daniel Lizars' daughter, Helen Hutcheson. Along the way, he became registrar of the United Counties of Huron, Bruce and Perth, a magistrate, and a collector of customs. According to some accounts, he was also named president of what became the Buffalo and Lake Huron Railway in 1856.[6]

Instructions to the Commissioner

On January 22, 1829, the Court of Directors had a very fulsome meeting with company solicitor James W. Freshfield in attendance. The meeting dealt with the "Special Commission to Mr. Jones to notify Mr. Galt of his being superceded and to receive all Property documents etc belonging to the Company," a "Deed of Covenant between the Company and Mr. Jones" and a set of "Instructions" addressed to each of Jones/Markland or Jones/Allan should Markland turn down the company's job offer.

Should Galt refuse to co-operate in the handover, Jones was to be instructed to use any lawful means at his disposal, including "one or more Attorney-at-Law … to appear for and represent the said Company in any Court or before any Magistrate as the case may require."[7]

The directors decreed that Jones' appointment was to be for one year, and that he was to proceed to Canada "forthwith." The "Instructions" for Jones and the other new commissioner dealt with a range of issues. Primarily, the new commissioners were called upon to review fourteen documents including the company prospectus under which the company was formed through to agreements with the British government and Lord Bathurst's instructions in 1825 to the five commissioners to determine fair value for the lands the company was seeking to obtain. Then there was the Royal Charter of Incorporation, the commission granted to John Galt and his instructions, the instructions issued to Thomas Smith, Thomas Smith's report and a series of extracts from Proceedings of the Court of Directors. Exchanges of correspondence with John Galt were also to be reviewed.

The documents and papers, said the directors "are recommended to your attentive perusal as affording a knowledge of many circumstances relative to which you will be called on to form opinions or to give decisions, and also as elucidating the opinions heretofore entertained, and the instructions from time to time given by this Court in regard to the management of the Company's concerns." … "It will be obvious," continued the directors, "that the first great object of this as of all similar associations must be to obtain adequate returns and an ultimate object for the Capital required in carrying on their operations."

This objective was to be reached by selling the land which the company had contracted to buy from His Majesty's Government at "enhanced prices." Revenues from land sales were to be used to meet company expenses including the cost of the land the company had contracted to purchase, salaries and expenses of their establishment, and improvements "whereby the value of the Land may be enhanced, and its settlement facilitated." The improvements issue, they noted, had always been the biggest source of difficulty and had been the subject of much discussion "during the management of the late Superintendent."[8]

To meet their stated goal, the directors stated that all expenditures at Guelph were to be stopped until the report they had commissioned on the state of the settlement had been received. In their view, many expenditures there had been "intrinsically unnecessary." Outstanding accounts, however, were to be paid. Expenditures were now to be limited to the Huron Tract lands and no improvements were to be authorized there other than for the opening of roads and construction of bridges. A special report was requested of the commissioners on the state of the Goderich settlement and on the new road from Guelph to Goderich. The directors then made note of the fact that the previous April, authority had been given "to lay out a Town at or near Goderich Harbour," with £3,000 made available "for the expense of forming that settlement and opening road access thereto." However, no report on the progress made nor expenses incurred had been received. The commissioners were instructed to prepare one and they were also to report to the directors on what else should be done for the improvement and settlement of the Tract.9

With regard to payments and receipts, the directors drew the commissioners' attention to the extracts of correspondence with which they had been provided. "You will see," they said, "how much dissatisfaction has arisen in consequence of the want of attention to this most important object." They emphasized that duplicate vouchers were to be used for all payments and receipts for the company account. One part was to stay at the company's head office in Canada and the other part was to be sent to the company office in London. At the beginning of each month, they were to dispatch a statement of all cash transactions for the preceding month. The statement was to specify the balance at the beginning of the month and the balance remaining at the end of the month, along with a statement of receipts and disbursements during the month.10

Although the directors were determined to have full control, there were certain aspects of the company operation which of necessity had to be left to the discretion of the commissioners. The determination of price and methods of sale were in this category, but the commissioners had guidelines. Their instructions read:

> ... it is not the policy of the Company to remain great holders of Land provided it can be disposed of at fair prices compared with the value of adjoining Land. Neither is it the intention of the Company to make sacrifices, or to force sales where such fair prices cannot be obtained.11

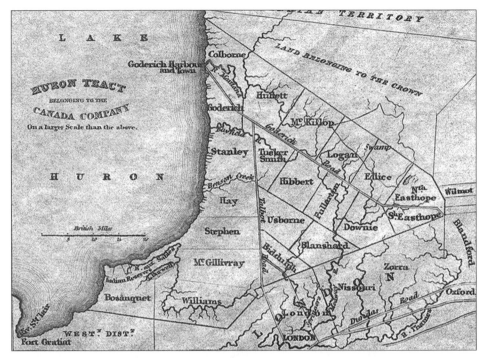

Map of the Huron Tract. Although it forms part of the Canada Company's 1841 advertising broadsheet, the map appears to be late 1827 or early 1828 in that the only town shown is Goderich which was founded in 1827. Note the Goderich (or later Huron) Road (now Highway #8) and the Talbot Road (now Highway #4) from present-day Clinton (originally The Corners) to London. *Courtesy of the Huron County Museum, John G. Goodwin Q.C. donation.*

Until the report commissioned by the directors was received as to the value of the various company lands, the commissioners could only accept offers for not less than twice the price paid. The directors, however, realized that they might have to accept a lower price in future "for inferior" land (they were obviously not aware of land values because they were trying to justify a price of 7/2 an acre which was just too high, given that "wild" land was selling for half that price at the time).

Galt's serious shortcomings were reflected in the instructions issued to the commissioners:

> Without regularity in Cash transactions and punctuality and
> promptitude in recording and transmitting the proper accounts

therewith connected no Concern can be satisfactorily managed, and by referring to the Extracts of Correspondence you will see how much dissatisfaction has arisen in consequence of the want of attention to this most important object.[12]

Although, from their perspective, the directors made it known that York was the most desirable location for the principal office, they said they would refrain from coming to any definite decision in this regard until a report on the subject had been submitted. One disadvantage was seen in moving the principal office from Guelph. The settlers might think the company was leaving the community to fend for itself, thus discouraging emigrants from settling there. This would, of course, affect future returns, but the directors were willing to take that risk.

Office expenses and salaries were the next item on the agenda. Again, the directors reiterated the need for "strict economy," given the present circumstances of the company. They discussed Dr. Dunlop's remuneration, and more generally the size and cost of the company's establishment in Canada. They noted that Dunlop's emoluments had been fixed "by a board of officers at Quebec" at a higher rate than had been contemplated by the directors. That magnitude, they said, prompts an enquiry as to whether it is necessary to continue his office. Having ascertained that his work was now chiefly in the Huron Tract, and as they had not received anything in the way of a report in the past year, they charged Jones with preparing a report "specifying the nature and the utility of the services heretofore rendered or heretofore to be expected in the Department called the Warden's Office."[13]

In reviewing the eight clerk positions which Galt had established in Canada, they left it to Jones' discretion as to what to do, pending a promised report from Thomas Smith. With regard to Charles Prior, "another person who has been employed by Mr. Galt and who had the immediate management of the operations at Guelph in 1827, where considerable sums of the Company's Money appear to have passed through his hands," they noted he had moved on to the Huron Tract in 1828. They hoped, they said, that "the example of Guelph would not be followed in regard to expenditure." Because the accounts had not yet been received, and because they did not know his rate of pay, they asked Jones for a report on this situation too – but they left it to him as to whether he wished Prior to stay in charge of the new settlement in Goderich. If in the affirmative, they suggested that he be required to give security to indemnify the company against the consequences of "malversation or disobedience of instructions."[14]

In concluding their instructions, the directors declared that it was absolutely necessary for the commissioners to be guided by them. Any deviation would not be tolerated:

> ... even if you should at any time form the opinion that the instructions transmitted to you have originated in mistakes or in misinformation, still you are not to act in disregard of them, but to report such cause as you may have to shew for any desired alterations and to wait the final decision of the Court thereon.[15]

Many factors made it necessary to practice a policy of strict economy and to make every possible reduction of expense. There had been the initial delay resulting from the commissioners' report of 1825 and the subsequent hold-up of the company's charter. Then, in 1826, came the exposure of many "bubble" speculations in South America as well as in the United States where shares were being run up away beyond their value by speculators. This was impairing public confidence in all joint stock companies, including the Canada Company. Public opinion was speculating that the company had overspent.

While the value of the stock was depreciating and company shareholders were becoming disenchanted, the directors were determined to put the company back on its feet.

Assessment and Conciliation

The new commissioners quickly began their task of assessing company operations in Upper Canada and making suggestions as to how to best proceed while negotiations as to the company's future were underway in England. It would not be long before the directors implemented change in Upper Canada. In line with the announced economy measures, but equally so because the directors learned that Tiger Dunlop was now becoming involved in local politics, and using this involvement as the hook they needed, they decided to discontinue the office of Warden of the Forests.[16] His political speeches, they said, "must certainly give offence to many persons ... whose goodwill might be beneficial, or whose displeasure might be injurious to the Company."[17] William Allan was

Louisa (McColl) Dunlop was housekeeper to the Dunlop brothers, Tiger and Robert. Bending to pressure from the community to marry, they flipped a coin to see who would marry her. Robert won (or lost). Thanks to the "Tiger," they had flipped a two-headed coin. *From* In the Days of the Canada Company *by Robina and Kathleen M. Lizars, 174.*

not in favour of this course of action and immediately wrote to the directors. Dunlop, he said, was the only person in their employ in Canada who had a complete knowledge of the company affairs, and could be extremely useful.[18] The directors listened, and Dunlop's dismissal was rescinded.[19] Soon after, the commissioners were reminded that employees were to devote their attention to company business exclusively:

> The Directors have to advert to the principle already acted on, that their Servants are not to engage in any public political matter or pursuits, or in any commercial or other objects, but that the whole of their time and attention shall be given exclusively to the affairs and interests of the Company.[20]

The directors' worst fears about the operation in Canada were confirmed as the commissioners dug deeper into the Guelph office. In 1827, a three-part form of sales contract had been drawn up by the directors to simplify operations. Part one was to be retained by the company agent or commissioner at his office, part two was to be given to the purchaser while part three was to be sent to Canada Company headquarters in Upper Canada for onward transmission to London. Title deeds would then be prepared for the buyers and sent to Canada. To their chagrin, Jones and Allan discovered that under John Galt's watch, no sales contracts had been sent to London between March 1827 and September 1828, in

This sketch of early Guelph, circa 1845, shows the Priory and the flour and gristmill on the Speed River. Originally the mill had been built by the Canada Company to serve the needs of the settlers. It was purchased by William Allan (originally from Ayrshire, Scotland) in 1832. He subsequently bought the Priory as a residence for himself and his family. *Courtesy of the Archives and Special Collections, McLaughlin Library, University of Guelph.*

spite of Galt having had eight clerks to assist him. Consequently, no deeds had been prepared.[21] Settlers were becoming increasingly upset with the company and were clamoring for their deeds. Jones also discovered that there was scarcely a settler who did not have a claim for wages over and above the amount owing on the next installment, and the accounts and receipts for land sold were in disarray. No regular set of books had been kept. Cash transactions had been particularly badly handled. Galt had "managed" them in his house, keeping a kind of account current with the company. He had used the account books in the office for petty transactions only. Jones doubted the book value of company houses. The Priory, for instance, in his words "handsome and commodious," could hardly be sold for anything approaching its cost.[22] In addition, the warden's office, which was being used for general company business, had been poorly built. The walls were cracking and had to be supported with large blocks of masonry. Instead of doing more useful work, such as arranging for the repair of the office, superintending the building of roads and bridges and dealing with squatters and illegal logging on company lands, Jones discovered that Dunlop and company employee Charles Prior had been working in the counting house for the past twelve months "doing what Lads for small salaries would have done equally well."[23]

Jones reported on, and reiterated, the problem with titles. Because buyers, and potential buyers, doubted the ability of the company to grant titles, both building and improvements were being held up. Many settlers were willing to pay the last instalment on their purchase, he said, but were waiting until they could obtain their deeds.[24] "This gives Mr. Robinson [in Peterborough] a great advantage over us …," stated Jones.[25] Allan, meanwhile, reported that a party of thirty families appeared lost to the company because of the uncertainty of obtaining deeds. He asked:

> – how many good and equally desirable Settlers therefore must
> the very same causes have lost us that we have never heard of?[26]

Following Galt was not going to be easy on many counts. He had become a favourite with the settlers and had done much for them, well beyond the scope of what he had be authorized to do and, in doing so, had become too much of a "people person." It now fell on Jones and Allan to stop all expenditures in Guelph, expenditures which the settlers took for granted. Seed, for instance, was all important. Galt had always advanced funds for their purchase and settlers expected this practice to continue. However, if advances for seed were not made, Jones reported in correspondence with the directors, settlers would be reduced to starvation the next winter.[27]

Because he was a realist on this issue, Jones advanced between £200 and £300 for the purchase of seed without authorization. A severe reprimand followed. In no uncertain terms the directors said they wanted it understood that there were to be no advances made, especially those to be repaid in labour as Galt had so often done. Talk of flourishing settlements or increased land values did not matter. Such advances could not be converted into the means of paying instalments, dividends or the expenses of management. It was put bluntly:

> The Court of Directors have not in any instance authorized or
> sanctioned the advance of money or money's worth to Settlers.[28]

Any doubt as to where the principal office should be in Canada was dispelled on reviewing John Galt's day-to-day activities; it was a precondition of Allan on becoming a commissioner that the office be moved to York in any case. Despite Galt's insistence that most business was done in Guelph, the new commissioners discovered that he had made frequent trips to the office at York on company affairs. This had resulted in delays in dealing with land applications,

Canada Company Headquarters, Toronto. The headquarters of the Canada
Company was moved from Guelph to York when William Allan became company
co-commissioner in 1829. The company moved into this building in 1834.
Owned by Allan, it was located on Frederick Street between Front and King.
Courtesy of the Toronto Public Library (TRL), J. Ross Robertson Collection: JRR 823.

and a loss of sales. In many cases, prospective purchasers went elsewhere having
concluded they could not complete their desired purchase. Because of poor
communication between the two offices and Galt's abysmal record keeping, it
was not unusual to have the same lot sold twice.[29]

If the company was going to keep afloat, sales – the lifeblood of the
company – would have to be increased. Accordingly, in June 1829, general
agents were appointed in various parts of the province[30] and special agents were
assigned to Quebec, Montreal and New York. The Quebec agent was to board
each immigrant ship upon its arrival. Armed with a company pamphlet, he was
to induce the new arrivals to settle in company territory. Dr. Dunlop was sent
to make all necessary arrangements with the new agents and was to report on
local circumstances which might enhance the value of the company land.

The new co-commissioners decided against offering lands at public or
auction sales. They had conferred with Peter Robinson, the commissioner of
Crown Lands, and had learned from his experience. At Crown Land sales,
seldom had more than a few lots been sold at a time.[31] While prices often went
higher at such sales, prospective purchasers were scared off because of this.

In a long letter to the Court of Directors, Jones gave a hint of the work
ahead and the money which would have to be spent in order to attract settlers.[33] He reported that the road from Guelph to Goderich was in a deplorable

state. He said it was merely a sixty-mile passage with branches meeting at the top. After a rain, it was like a bog and took days to dry because the sun could not get through. He recounted that on an inspection tour over the road, "or rather through it," the mud was frequently up to the horse's belly. Accordingly, Jones recommended that it be cut to its full intended width of four rods (66 feet) in order to alleviate this condition. He estimated the cost of doing this at £1,250. It could be done at this comparatively cheap price, he thought, because only those agreeing to be paid five-eighths of their earnings in land would be employed. He also suggested that other roads be opened, but owing to the unsettled state of the company's affairs, talked only of the road from present-day Clinton to the Talbot Settlement near present-day London. This would facilitate communication and settlement from the southern end of the Tract. He recommended that roads should be opened in the Town of Goderich and noted that considerable expense would be required to preserve the harbour there. He held out some hope, though, that the government would make the town a military post. This would mean that certain improvements would be borne by the government.

The harbour at Goderich was very important to the company. It was the best facility on the Canadian side of Lake Huron through which to conduct trade and receive settlers. While its location at the mouth of the Maitland River should have made for a fine natural harbour, in fact, because of a fairly broad sandy estuary, limited depth and a western exposure, strong northwest winds kept shifting the sand. Sandbars were the inevitable result which meant that keeping the harbour open was a constant drain on company finances. In fact, in 1835, it was closed for the whole of the shipping season.

The directors were caught in the horns of a dilemma. The shareholders were demanding a change in the agreement with the government and no decision had yet been made as to what to do. The measures suggested by William Allan and Thomas Mercer Jones were obviously necessary, but the directors were not inclined to making further funds available. They still had no complete report of what the company property included. Further, they wondered if the existence of the company could be prolonged at all. In the interim, the best they felt that could be done would for the commissioners to obtain estimates for constructing a mill dam and erecting a gristmill, sawmill and fullingmill (used in wool-making process) at Guelph. They were also to ascertain the rent which could be obtained for these mills, or if the water power could be leased.[34] If the company were to retain the Huron Tract, the opening of the two main roads would be essential, but from the director's perspective, the cost should come out of the Huron Tract Improvement Fund. To this end, the commissioners were instructed to ascertain the views of the government of Upper Canada on this matter,[35] and this became a major task to be pursued in earnest.

From the end of August 1829 until the meeting of the proprietors in the London Tavern, (their usual meeting place[36]) on January 7, 1830, neither the directors nor the commissioners could predict the future. Indeed, in a letter to the court, Jones reported that they were suspending operations in Canada pending something more favourable from London.[37] While both commissioners were convinced of the value of the Canadian property, William Allan wrote to the directors that shareholders must be patient:

> We conceive the Land well worth much beyond the Purchase Money, but it cannot be regarded as an article whereby a speedy return of Capital is to be expected or looked for. It must be looked upon as rather forming a permanent Stock yielding for the present an annual interest in the shape of dividends and to be returned eventually with enhanced profit.[38]

To be realistic, the commissioners saw that the very nature of the project, in the short to medium term, could not yield sufficient funds to sustain payments to government, meet the contingencies of management and provide a profit to be divided. As far as Allan could see, the difficulties of the shareholders had arisen from two causes: a mistaken view of the subject and nature of the speculation, and a hesitation in advancing the necessary capital.[39] Regular calls should have been made on the shareholders for capital still owing on the shares subscribed, in order to keep to the terms with the government and to assist in meeting company expenses. The receipts, or a sufficient portion of them, should then have formed a fund as profit. From this profit and the monies derived from sales, an annual half-yearly dividend could then have been paid to shareholders. Given this scenario, the amount realized from sales to September 1829 (£12,771/1/7 currency)[40] would have yielded about six per cent on the actual calls from the shareholders, and would have supported the value of the stock in the market.

Commissioner Allan continued that the number one objective in the Huron Tract should be the building of roads and the erection of mills. Roads were necessary for access, and it should be considered essential to provide enough capital each year to keep them in good repair. Mills and roads, he pointed out, enhance the value of adjacent lands. He warned, however, that persons with capital initially would not be willing to support the erection of mills. To overcome this, he cited the Peter Robinson's settlement, which became Peterborough, as an example. There, a gristmill had been built simultaneously, at government expense, with the arrival of 1,825 dispossessed Irish tenant farmers and their families. When the settlement was deemed successful, the mill and site were then sold at auction, at a price not exceeding the cost.

According to Allan, it was important to encourage pauper immigrants and to adopt an arrangement for taking produce. He also suggested that two offices should be established for local management. One would deal exclusively with the sale of the Crown Reserves, to be paid in cash. Since no improvements would be undertaken on them, little supervision would be required. The other office would take total charge of the Huron Tract lands, where improvements would be required. Separate books, accounts and clerks would be used. The reserves, he said, would appeal to friends and relatives of the earlier inhabitants of the old settlements and would sell at a higher rate. The Huron Tract, a more remote settlement, would suit any number of families who might desire to form a settlement together. While the Tract would yield the greatest profit in proportion to the original cost in the long run, the reserves, he felt, would yield the readiest and earliest returns.

He stipulated that a key duty of the commissioners in Canada should be to press for interest and installments as they fell due. This had not been done and he expected that an allowance of 25 per cent would have to be made for bad and doubtful debts. He stressed the necessity of the company being prompt in its payments to government and in its dealings with purchasers. Confidence had to be restored.

Having addressed his report to Simon McGillivray as chairman of the Committee of Correspondence, Allan then wrote a letter to him personally. In it, he apologized for making what were perhaps "improper or impertinent remarks," but said he was rather unaccustomed to making reports such as the one he had presented. He added that he hoped it would be taken in the spirit intended, i.e., as constructive criticism and suggestion.[41]

While Allan and Jones were working to straighten out Canada Company matters in Upper Canada, Simon McGillivray would have received Allan's letter at a time when he, McGillivray, had become disenchanted with the company's shareholders and frustrated with his fellow directors. So much so, in fact, that he officially tendered his resignation at a meeting of the Court of Directors on September 3, 1829, not only as chairman of the Committee of Correspondence, but as a company director as well.

When he agreed to take on the chairmanship of the committee in April 1827, McGillivray reminded his fellow directors that his doing so would be an experiment, and that he was only prepared to remain at the helm of the committee at the pleasure of the directors and in consideration of his own arrangements. In his letter of resignation, McGillivray emphasized that he did not want to be seen as disserting his post nor indeed of being perceived as abandoning "a falling House."[42]

McGillivray's £1,000 a year salary was the issue that forced the resignation. Company shareholders had been arguing over this salary at both their January and March meetings that year. To McGillivray, it was an intensely personal matter. It was evident, he said that "the object was to represent my Salary as more burdensome than my services were beneficial to the Company." He felt he had no choice but to resign.

The resignation was accepted with sincere regret, with the directors noting the "diligence and talent" which been exhibited by their colleague. Because they knew no other member of the Court possessed the same level of knowledge about the business of the company and because none had had a comparable level of experience in Canadian affairs, the directors would now be forced to change their *modus operandi*. A much more "hands on" approach would be required, with more frequent meetings of Court the norm.

As a result of McGillivray's resignation, the directors made it known quite frankly that those directors who were not able to devote an appropriate amount of time to company business "especially under the present circumstances of embarrassment and difficulty" should resign. At the same time, they invited shareholders who possessed knowledge of Canada, and undoubtedly the requisite number of shares (see Appendix B for more on the directors), to join the court because those few directors who had knowledge of things Canadian seldom attended regular court meetings.

While the directors were clearly going to miss McGillivray and his expertise, the former committee chair and director must have been disappointed that his colleagues were not prepared "to push back" and support him on the salary issue at those last two meetings of shareholders.

Having dealt with the McGillivray resignation, the directors turned to the matter of a permanent replacement for Galt as company secretary. A.S. Price, who had been doing the job in an acting capacity, was appointed at a yearly salary of £400.[43] With important housekeeping items dealt with, the directors would now focus on trying to keep the company afloat.

William Allan's report to Simon McGillivray made a considerable and positive impression. Many of his suggestions were put into practice and, with a solid relationship developing between the commissioners and directors, the tight rein from London began to loosen. The new commissioners were granted permanent powers of attorney to execute deeds on all contracts, but from this point on, no title was to be given until the whole purchase money had been paid.

Thomas Mercer Jones worked particularly long hours trying to unravel the accounts left by John Galt. On a daily basis, undocumented invoices were being presented for work or purchases authorized by him. To Jones and Allan, many were unnecessary, but they all had to be paid because Galt had authorized them. Allan stated: "The spending of Money as long as it could be placed against the Company seems to have been no consideration with any person heretofore in their employ." He continued, "We have laboured hard ourselves daily from 9 to 5 or 6 o'clock, even copying our own Letters, but this is rather too much of a good thing, and more I am sure than the Company wish; as to Mr. Jones he is hard at work on Sunday as well as on the Week days."[44]

While the commissioners were working hard in Canada to improve the company's image, former accountant Thomas Smith seemingly was attempting to sink the company. He published an address to the shareholders in 1829 which attacked both directors and commissioners.[45] Among other accusations, he charged Allan and Jones with paying out at least £750 more in accounts than would have been necessary had they consulted him. The highly respected William Allan was not pleased. He retorted to the directors that not "one shilling" had been paid which was not "fairly and fully proved to be due to the parties." He lamented that such a publication should appear at all, but particularly at this crucial time. He was afraid it would be reprinted "again and again":

some will make use of it, as a new bug bear to frighten People
from having to do with the Company – What can possess any
person who has had access to the business and transactions of
any Concern to make such uses of it![46]

Jones later wrote that had Smith devoted only one-twentieth of the time he
spoke of to the legitimate duties of his office, "our present labors would have
been much less and much of the confusion in the Books and Accounts would
have been avoided."[47] The overall state of the company's management in
Canada was summed up by Allan:

your whole Concern seems from beginning to end, to be a
most unfortunate one from the way it has been managed and
from the kind of people that has had to do with it.[48]

Once again William Allan's comments made an impression on the directors.
As a bank president, member of the Executive Council and businessman, he
was respected. If he was gloomy about the state of past management, he was
optimistic about the value of the company lands:

The Stock holders little know the value of what they possess,
or they would not hesitate a minute in deciding to retain it. We
have reason to believe that the persons at the head of affairs in
this Country are so fully satisfied at the immense bargain the
Company have got, they would most gladly hear they have
determined to give it up.[49]

The directors themselves knew this, but confidence had to be restored. It would
now be important for the commissioners to concentrate on land sales and collections.

Survival of the Company

As soon as the proprietors approved of the continuation of the company in
January 1830, the directors were in a position to act on the reports supplied by
the commissioners, and to implement many of their suggestions. As early as

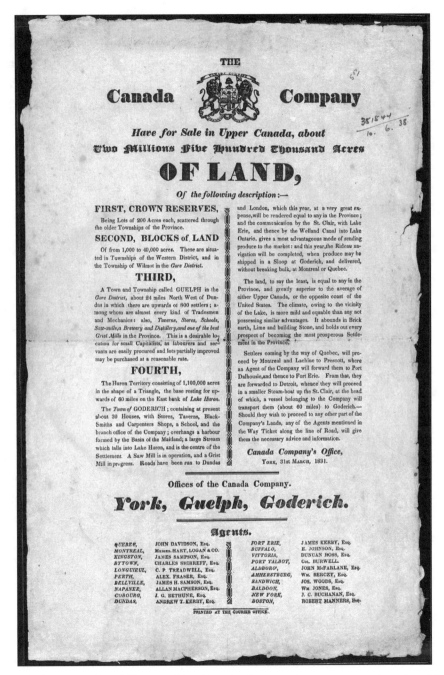

Canada Company Handbill, 1831. The Canada Company began an extensive advertising
campaign once the shareholders signified their continued support for the company.
By 1831, Guelph had "upwards of 800 settlers" while Goderich contained "about 30
Houses…" *Courtesy of the Thomas Fisher Rare Book Library, University of Toronto.*

1827, agents had been appointed in the outports of the United Kingdom and the ports of debarkation in the United States, as well as on the route from these ports to York in Upper Canada. The duties of the agents in Britain had been to show maps, diagrams and the company prospectus to potential emigrants. The agents at the ports of debarkation were responsible for arranging transport to Upper Canada and were to obtain supplies for settlers. The directors determined to complement their work by carrying out an extensive scheme of advertising in the British Isles particularly and, to a lesser extent, in the United States and Canada. An updated prospectus was prepared and bills were printed outlining particulars of the land and facilities offered to the emigrants by the agents. Attached to the prospectus was a map of Upper Canada showing the location of the company lands. In addition, there was a description of climate, soil and crop production that could be expected.

The prospectus and bills were posted and circulated in all the market towns and villages of the United Kingdom, and translated versions were posted at Le Havre in France. Circulated along with the prospectus were letters from satisfied settlers. A great deal of stock was placed in these letters because these would be received with more confidence than company advertising. Of course, it is important to note that only letters favourable to the company were published! A company agent, William Cattermole,[50] was employed to visit the various parts of the United Kingdom, including the Isle of Wight and the Portsmouth area. His job was to distribute company literature and answer questions about the colonies and, more specifically, about the company's lands. The publicity, of course, not only aided the company, but also was of great value to the colony as a whole. For example, English-born James Cull[51] (engineer, businessman, newspaperman and surveyor) was probably among the "persons of Capital" who accompanied Cattermole to Upper Canada in 1832. While Cull, who came with technical expertise as an engineer, did not settle on Canada Company land, nor was he an employee of the company, his road building expertise was put to work in the colony; he built the first macadamized road outside of York.

The company was realistic in its approach. The directors doubted that any settlers would enter into an agreement to take up lands until they were actually in the province. Immigrants preferred to be perfectly free to choose for themselves. Pressure from the company would only arouse suspicion. For this reason, it was most necessary to have the company agents in Quebec City, Montreal and New York. Since the prospectus was being distributed to the settlers as they embarked for the colonies, the newcomers would at least be aware of the company when they arrived. As a further encouragement to settlement, assistance in the form of a free conveyance was provided from Quebec City and Montreal to the company's lands from 1830 to 1833, but with certain stipulations.[52] To stimulate emigration northward from the United States, the company advertised in the newspapers of such cities as Albany, Utica, New York, Philadelphia and Pittsburgh.[53] Advertisements were also placed in Canadian papers.

With such extensive promotion and the hoped-for influx of emigrants, it was necessary to think seriously of how to transport the would-be settlers to company lands, especially the Huron Tract. Originally, the directors had thought the best way of getting settlers there was by way of a road running from York through Guelph to Goderich. However, as a result of the commissioners' recommendations in 1830 for the establishment of a regular link by water from Lake Erie to Goderich, Jones and Allan were instructed to make arrangements for chartering, or as a last resort, purchasing, a ship.[54] The directors had not wanted to become shipowners because of the responsibility involved. Unfortunately for the company, no other means than building and operating a ship was available and so, in 1831, an ill-fated scheme to transport emigrants by water began. For the first two years, the company had a schooner, the *Lady Goderich*. In 1833, the keel of the steamboat *Menesetunk* was laid. The directors became so convinced of the value of the steamboat that they considered it to be of equal importance to the overland stage route, which had been established in 1832 to run between Hamilton and Goderich. They reported to the shareholders that although the expense of running the boat would be high, the results would more than warrant it. They foresaw a large increase in the number of settlers arriving and a corresponding increase in the value of company land. Indeed, the price of land was raised on the launching of the *Menesetunk*.[55] But few emigrants arrived by ship. She ran at a loss continually, most of her freight being timber/lumber resulting from a ill-conceived scheme of Jones. The inglorious career of the *Menesetunk* ended in 1839 when the steamboat collided with another ship on the Detroit River.

The Goderich harbour proved just as troublesome. Having agreed with the commissioners that the future prosperity of the town and surrounding area depended on it, the directors were committed to its upkeep. Before finally relinquishing control in 1859, by happily selling it to the Buffalo and Lake Huron Railway for £13,000, the company was to spend nearly £17,000 for maintenance

Admiral Henry Wolsey Bayfield, RN. Largely self-trained, this noted maritime surveyor was only 22 when he began the hydrographic surveys of Lakes Erie, Huron and Superior (1817 to 1825). In 1827, at the request the Belgian nobleman, Baron de Tuyll, Bayfield identified a 4,000 acre site 12 miles south of Goderich which became the Town of Bayfield at the mouth of the Bayfield River.
Toronto Public Library (TRL), J. Ross Robertson, Collection: JRR 1666.

and repairs. The harbour issue was to develop into a constant source of friction between the directors and commissioners on the one hand, and the Huron Tract residents, more particularly the Colborne Clique, on the other.

Although the steamboat, harbour and reformer agitation were problems, there were also promising signs. Well-to-do settler Michael "Feltie" Fisher had come from Vaughan Township near York in 1831, and purchased some 6,000 acres in what is now the Benmiller area. In 1832, Belgian-born Baron de Tuyll[56] had purchased a total of 4,000 acres at the mouth of the Bayfield River overlooking Lake Huron for 3s.9d.per acre on the advice of hydrographic surveyor Henry Wolsey Bayfield.[57] The Baron thought he had found the perfect spot to establish a city.

By now, mills had been built at Guelph and Goderich, and accounts for 1831 were showing a balance of £50,600 in favour of the company on its operations.[58] The level of emigration remained high and the directors were wondering if it would be feasible to raise the price of land.[59] "The Corners" (later Clinton), twelve miles east of Goderich, had been laid out in 1832 at the intersection of the Huron (or the Goderich) Road and the Talbot Road to London. Further east, on the Huron Road, the community on the Little Thames which had been established in 1828 was named Stratford in 1832 by Thomas Mercer Jones – and the name of the river was changed to the Avon. More than satisfied with the new co-commissioners, they reported to the shareholders:

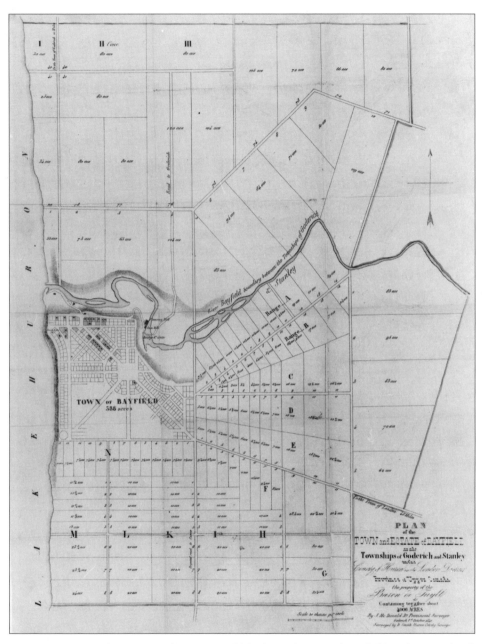

In 1830, provincial surveyor John McDonald laid out the site of Bayfield at the mouth of the river which Admiral Bayfield had recommended to Baron de Tuyll. Despite great plans, Bayfield consisted of only a few huts and a store six years later The Baron returned to Belgium in 1835 and died that year. His son took over but lacked the funds to develop the site. *Courtesy of the Archives of Ontario, F 129 TRHT #234 AO 6487.*

The Town of Clinton was laid out by the Canada Company in 1831. It was originally known as "The Corners." This circa 1871 photo was taken at the intersection of what is now Highways #4 and # 8. *Courtesy of Kelvin Jervis, Jervis Photo Inc.*

> the Commissioners are regular and punctual in their advices and transmission of their accounts, and fulfil all their duties in a most satisfactory manner.[60]

This resulted in a further loosening of the directors' hold from London. The commissioners were allowed to use discretion in deviating from the usual system of land payment, ordinarily one-fifth down and the balance over five years:

> Particular cases and peculiar circumstances will always occur, which would make undeviating rules as to the amount of the first payment and the extent of credit inconvenient, if not injurious to the interests of the Company, and therefore, the Directors leave you to the exercise of the necessary discretion therein.[61]

With success came the idea of purchasing further territory. Specifically, the company wished to acquire lands (now Puslinch Township) adjoining the Halton Block which at the time was a Clergy Reserve. Interest was also expressed in lands north of the Huron Tract (more specifically north of Colborne Township),

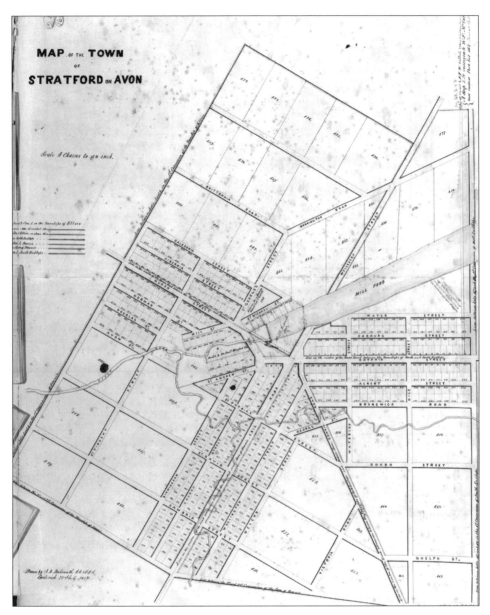

Map of Part of the Town of Stratford. Settlement of the site began with the
surveying of the Huron Road in 1828. Canada Company commissioner
Thomas Mercer Jones named the community Stratford circa 1832 and renamed the
Little Thames River, the Avon River. The town was laid out by Surveyor
John McDonald in 1834. It was incorporated as a village in 1854 and a town in 1859.
Courtesy of the Archives of Ontario, F 129 ATLAS 11, 1517, #24, AO 6479.

not only as an outright purchase, but also in case of a deficiency in the Huron Tract grant which may have to be settled. The directors noted:

> – it will ... be the more incumbent on the Directors to manage their affairs with great caution – policy will induce them to sell freely on a good market, so that they may be able to hold off upon one of a contrary description, and also then be in a condition to buy, if it can be done advantageously without embarrassing the Company's funds.[62]

Their idea to purchase additional land was quietly withdrawn when the sentiments of the British government were learned. The government knew that the company had obtained a very good bargain and this knowledge was re-enforced by Lieutenant Governor Sir John Colborne. The directors decided to leave well enough alone.

As the company became better known, business promoters in the colony wanted the Canada Company to purchase stock in their enterprises. Such companies as the Welland Canal Company, steamboat companies on Lake Simcoe and on Lake Ontario, and the Grand River Navigation Company[63] (financed primarily by the Six Nations Indian band to improve transportation on the Grand River in order to give area inhabitants an inexpensive outlet to markets[64]) all tried to attract the Canada Company's interest, or rather, its capital. It was perhaps fortunate that the company, by the terms of its charter, was prohibited from investing in any joint stock company.

The Company Under Attack

As long as land was being sold, accounts paid, mills built and roads opened, no one – government, settlers or directors – had cause to complain. The serious cholera outbreak in 1832, however, slowed emigration after that year and arrears in payments totalled some £25,000 by September 1833.[65] The directors were forced to take a harder line and pressed the commissioners to collect arrears. The company, through its commissioners, became the butt of abuse in Canada. In England, meanwhile, despite record land sales of 114,804 acres in 1832, the company was once again coming under attack and the price of their shares was fluctuating on the market.

William Lyon Mackenzie was a leader in the Reform movement. Through his newspaper, the *Colonial Advocate,* he kept chastising the Family Compact and the Canada Company – and precipitated John Galt's first run-in with Lieutenant-Governor Maitland. Following the collapse of the Rebellion of 1837, he escaped to the United States. Eventually pardoned, he returned to Canada in 1849. *Courtesy of Library and Archives Canada / C-024937.*

The directors' fear of losing shareholder support became evident after the publication of a pamphlet by an anonymous writer who warned investors to be extremely cautious if purchasing company stock.[66] The writer said he had bought a quantity of shares at £50 per share (although only £17 per share had been paid in at that time). He had purchased them, he said, with the "fallacious idea" that they would be a good investment. A short time later, the stock dropped seven or eight pounds per share in one day, and becoming nervous, he had sold out. He warned: "The market is again advancing in price. Be cautious you do not fall into the same error."[67] Politics, war with the United States, separation from England or other difficulties may take place and he predicted increased competition from the sale of government Crown Lands. He lamented the low interest rate paid by the company (which was 4% at the time). In England, he stated, an investor could obtain 41/2% to 5% on a mortgage and have the principal returned.

In Canada, the newspapers were beginning to take sides. The Toronto *Correspondent and Advocate,*[68] for instance, was a warm supporter of William Lyon Mackenzie, who became the first mayor of Toronto in 1834. Edited by Father O'Grady, a suspended Roman Catholic priest, the paper was violently opposed to the Family Compact and the company. the *Patriot and Farmers Monitor,* on the other hand, was a fiery Conservative paper, supporting the Family Compact, the British connection and the company.

The Correspondent and Advocate sought every opportunity to discredit the company. A good example of this was the application by the commissioners to the legislature for an act authorizing the company to levy tolls on vessels entering the harbour at Goderich. The object of the tolls was to help offset the extensive expenditure required on the harbour to make it safe for ships to enter. The paper exclaimed:

> To such application we could have no objection, if made with a view to the public interests. But this is wide of the Commissioners' object. The aggrandizement of the Company at the public expense is their aim and their end.[69]

The *Patriot and Farmers Monitor* was more favourable to the company. One edition carried a copy of an editorial which had appeared in the Hamilton *Western Mercury*. It brought forward several facts which showed how much the company had contributed to the prosperity of Upper Canada, while at the same time proving profitable for the investors.[70]

Editorial comment and anonymous letters were not the only way of getting "at" the company. The Family Compact, that group of influential high office holders who, according to charges by the radical reformer William Lyon Mackenzie, achieved their social and economic objectives through control of the banks and land speculation, was difficult to attack directly. The company proved to be a very convenient scapegoat. Were not the two almost synonymous? The close ties between the company and the Executive Council were also resented by the Reformers, and so was the very idea of the company. The reformer-instigated *Seventh Report from the Committee on Grievances* not only attacked the existence of the company but also launched into what they thought was to be the inequality of taxation as it applied to company lands.[71]

> ... the industry of the country is highly taxed while the Canada Company reserves escape taxation.[72]

The Correspondent and Advocate reported in August 1835:

> ... disgust at Land Companies is not confined to Lower Canada. One of the last acts of the Assembly of Upper Canada,

previous to the close of its session, was to pass a resolution against the Land Company of that province. The resolution declares it to be "Unconstitutional in the British Parliament or Government to lay any tax on that province, or to dispose of any revenue arising therein, unless with the consent of the Assembly, whether such revenue arise from the sale of public lands or from district taxation." The Resolution, further, characterizes the sale of land to the Company as an improvident transaction, which "as it is unsanctioned by any domestic enactment, ought to be held invalid ...[73]

The Reformers resented the fact that the Assembly had no say in the disposal of the purchase monies paid to the colonial government for the company lands. This they viewed as a flagrant violation of the British Act of Parliament (18 Geo.11.c. 12)[74] which guaranteed that the colonies would not be taxed by the Imperial Legislature, and that all proceeds from taxation raised within the colonies would be placed at the disposal of the colonial legislature.[75]

The company was now coming in for criticism on all sides. Patrick Sherriff, for instance, a Scot who visited the province in 1833, was critical of the terms under which the land was sold. His view was that the terms were *too* generous and that they would result in settlers being tempted to buy more land than they could properly use![76]

By 1835 the Colborne Clique was becoming vocal, and Jones was asked by the directors to make more frequent trips to the Huron Tract in order to try to keep things under control.[77] The Colbornites were angry at the lack of what they viewed as a proper bridge over the Maitland River from Goderich to their township. One suspects that had they been less arrogant and more co-operative with the commissioners, an amicable arrangement could have been reached sooner. It would, however, be 1839, and after much wrangling on both sides, before a decent bridge was built. In dealing with the Colbornites, what the directors had failed to take into account, of course, was that of all the Huron settlers the Colborne Clique were among the most wealthy. They had a better than average education and had time to badger the company. The delay in building a suitable bridge only added fuel to the already smoldering fire of discontent directed against the company. From the beginning there had been a ferry boat of sorts and then a floating bridge from the Goderich Harbour across

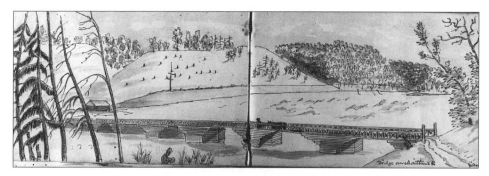

First Bridge over the Maitland River, Goderich. This bridge was built by the
Canada Company in 1839 after considerable agitation by the Colborne Clique. It
replaced firstly a ferry and then a moveable floating causeway. Because most lands had
been sold in Colborne Township by this time, the company felt the Colbornites should
be responsible for building the bridge. *Courtesy of Library and Archives Canada / C 098789.*

the river to the township, and since the river upstream could be forded at most
seasons of the year, the Colborne settlers were not quite as badly off as one is
lead to believe. But in order to bridge the river properly, a substantial and costly
structure was required. That is where the problem began.

The directors realized the need, but were hesitating. The thought of bene-
fiting Crown Lands seven miles to the north came to mind in that they did not
want to encourage settlement north of Colborne Township until any claim for
further land due to deficiencies in the Huron Tract had been settled. Secondly,
Colborne Township was off by itself and most of the Canada Company land
there had already been sold. It would be much more in their interest, the directors thought, to build a bridge in a more central location of the Tract.[78] They
reasoned that if the Colbornites really wanted a substantial bridge, they should
help defray the cost.

A further source of friction was the "tuppeny mill down on the Flats"[79] at
Goderich, built by company agent Charles Prior. It did not always run because
of an inadequate water supply, and pressure was being brought to bear on the
company for a more reliable mill. In the meantime, the directors were delaying
approving its construction in the hope that one or more inhabitants would
come forward to build one. No one did, and so the company was forced to give
in to the demands. In 1841, they built a substantial mill on the Maitland River
just east of Goderich, but in the early days, it only ran one or two days a week
because an inadequate supply of wheat.

Van Egmond House, Egmondville, as it looks today. There is still debate over whether Anthony Van Egmond was an imposter or a colonel and an aristocrat of some stature from his days in the Napoleonic Wars and before. He made a positive contribution to the Huron Tract by opening and building roads. The company paid him primarily in land, which annoyed him. Van Egmond supported the Rebellion of 1837. *Courtesy of Paul Carroll.*

The Colbornites were not the only source of irritation for the commissioners. Former ally of the company and friend of John Galt, Anthony Van Egmond, who had moved from his original farm at Oxford East to the Huron Tract and had purchased land at Egmondville, near present-day Seaforth, proceeded to organize and lead the Huron Union Society in 1835. Van Egmond had been the contractor who opened the Huron Road in 1828. He had little use for Thomas Mercer Jones because of his connections with the Family Compact, and became disillusioned over what he said were the neglected duties and broken promises of the company. Furthermore, he was annoyed at having been paid for two-thirds of his road construction work in land credits, instead of cash.[80] The aims of the Society included: a demand for responsible government, sale of the Clergy and Crown reserves by government, control of public revenue by elected officials and the abolition of all licensed monopolies, notably the Canada Company.[81] Van Egmond sent a copy of the Society's proceedings and resolutions of the first meetings to Canada Company directors. The reply from John Perry, company secretary, was a model of diplomacy, reflecting the company's interest in keeping the peace, maintaining shareholder support and, it was hoped, paying a dividend:

> I am to request that you will communicate to that Society this acknowledgement, and to express the sincere wishes of the Directors ... that all their plans and operations may be governed by considerate aim to promote in the District individual comfort and general harmony.
>
> The interests of the Canada Company viewed through a proper medium, can only appear to be identical with those of the Settlers upon their Lands; & the Directors in England, and their Commissioners in Canada, are deeply anxious to promote the comfort and prosperity of the Inhabitants of the Huron Tract.[82]

The company had many reasons to be diplomatic. Emigrant landings at Quebec had gone from a high of 51,422 in 1832 to a low of 11,539 for the first ten months of 1835.[83] The cholera epidemic of the two previous years was taking its toll on the colony more generally, and on company sales more particularly. Not only did the epidemic cause a fall in emigration, but the agricultural and manufacturing districts of the United Kingdom were in a less depressed state than formerly. Fewer persons were being induced to leave for the colonies. The political state of the Canadas was having a negative impact as well. William Lyon Mackenzie was agitating for change through his newspaper and the Reformers (generally people from the countryside) were arguing that the ruling clique (read Family Compact) was selfish and unrepresentative. They pointed to defects in the system of granting land, the inadequacy of schools and bad roads.

Thomas Mercer Jones gloomily reported on the state of affairs to the directors, including the greatly reduced emigration, the failure of the Harbour Bill in the Assembly which would have enabled the company to levy tolls on ships using the Goderich Harbour, and the land-locked condition of the harbour.[84] The company steamship, he said, could not get over the sandbar at the harbour entrance and three schooners had been wrecked trying to do so. Because of this, the price of flour rose had risen from $4.50/$5.00 a barrel to a high of $8.00/$9.00. The company was blamed.[85]

The directors were upset at being unable to levy tolls at the harbour. While acknowledging that it must be kept in a state of repair, they wondered why the company should continue to be solely responsible for its upkeep when so much of the land in the vicinity of Goderich had already been sold. Annoyance at the company had been the reason the Harbour Bill had been thrown out of the legislature and so, in a fit of pique, the directors proposed a solution:

> Let the great Proprietors near Goderich form a Company, and let the application for a Bill be made by them: the Canada Company might subscribe to the Stock, if they are duly empowered, to any extent they think right: the object would then be carried on, as in all similar cases, by Directors chosen to manage the undertaking, This appears now to be the proper course …[86]

This solution, predictably, fell on deaf ears and the *Menesetunk* remained landlocked throughout the summer of 1835. With the harbour clearly still the responsibility of the company, there seemed no alternative but to petition the Governor-in-Council for a lease of the entire harbour area. This would at least allow the levying of wharfage on all goods shipped or landed. Approval of the harbour lease was granted, but because the Maitland River was viewed as navigable water, a portion of the harbour was not included and so the company still did not have absolute control. Because of this, it was decided by the directors to spend only as much as would allow their boat to cross over the sandbar. But if their boat could cross the sandbar, so could others!

The company had been outsmarted by the Executive Council. Goderich was the only refuge for ships from Sarnia to Penetanguishene. If the company could not have been induced to maintain its portion of the harbour, the government would have been forced to carry out maintenance before long. The leasing plan had been an easy way out.

In exerting pressure on the directors for a boat (the sailing ship and then the steamboat) and harbour improvements, it was Jones' view that if a regular steamboat communication were to be established between Goderich and Lake Erie, "the Company may laugh at the threats of going to Michigan, which dissatisfied people are so frequently holding out."[87] The allusion to "threats of going to Michigan" no doubt arose from comments made by Hugh Black to the Honourable Robert B. Sullivan, the commissioner of Crown Lands. Black, as Clergy Reserve lands deputy-commissioner, had been on a fact-finding tour and had reported to Sullivan the adverse influence of the Canada Company as he saw it. From accounts he had received, it appeared as if very few emigrants with capital would be settling in the Western and London districts:

> The glaring accounts circulated in Britain by the Canada Company induced a number to settle there for these two or three years past, & now find superior advantage in locating themselves in Indiana, Illinois & Michagan & are now leaving the Company's lands by wholesale.[88]

Many who had already left were "doing their utmost" to get others to follow. He reported that because of "universal antipathy" which the settlers have towards the under-agents of the company, extensive correspondence was taking place advising friends in Britain to go to the United States. Settlers were being cautioned to locate somewhere between Toronto and the Grand River if they wished to stay in Canada, but "by all means to keep from the Company's land."[89] Black reported the belief among emigrants that the company land "patents" (deeds) were defective. He said his informants were not easily biased and would have no interest in misleading him:

> I can therefore assure you that the prospect of settling the London & Western Districts by emigration at present is some-what gloomy, the pleasing accounts in the newspapers notwith-standing.[90]

He continued his attack on the company as it related to the Guelph settle-ment. The land was "depreciating daily" from the removal of the Canada Company's "extensive establishment which always circulated some money through the immediate neighbourhood." What he termed "a most inconsiderate grant" was made to a Mr. Strange by the company for a mill seat (apparently it was the only one in the township with the exception of a falls near the town line). Strange, he reported, did not have the funds to develop it, but would not sell it either. This grant, said Black, depreciated the Clergy Reserves in the township as well as the surrounding areas "to the amount of £15,000 at least" and many settlers were forced to go to mills in other townships because of this.[91]

It seemed that the company could do no right and the commissioners were once again caught in the middle. In 1835, land sales of 58,751 acres were half

Captain Robert Dunlop captained the Canada Company ship, the *Menesetunk*, from 1833 to 1836. He had a falling out with the company and was relieved of his command. He became the first elected member of the Legislature for the Huron District in 1835. He died in 1841 and his brother Tiger ran in his place and won the disputed election of 1841. *From Mrs. W. Webster, Montreal, and publisher of Dunlop, J.C., Dunlops of Dunlop and Auchenskaith, Keppoch and Garbriad. London: Butler and Tanner, 1939.*

of the 1832 total. Perhaps certain complaints against the company were justified, but whether they were or not, as of June 1835, the Colborne Clique had a spokesman to air their grievances in the Assembly. That year, Captain Robert Graham Dunlop, Tiger Dunlop's brother and a Colbornite, became the first elected member of the Assembly for Huron, defeating Anthony Van Egmond, fifty-nine votes to two.[92] An anomalous situation was developing. Robert became the spokesman for the Colborne Clique against the commissioners and the company, while in 1836, Tiger was writing his *Defence of the Canada Company*. In the midst of all this, Charles Prior was dismissed by Jones for misappropriating company funds.[93]

While much of the criticism being voiced against the company was actually meant for the government, the company continued to be a convenient scapegoat.

Company Policy Reinterpreted

Thomas Mercer Jones and William Allan, knowing of the building anti-company mood, made some timely suggestions to the directors early in 1837 in an effort to deal with it.

To address many of the grievances against the Canada Company, an expenditure of money was required since the complaints centred around the opening/building/upkeep of roads and bridges and the building of mills. The directors did not deny the need for this, nor did they have any doubt of their power or ability to raise capital, but they were only willing to do so if the advantages of proposed projects could be easily ascertained, and if the funds required could be clearly defined when undertaken:

> The Board are free to confess to you, that the principal motives which have restrained them in any appropriation of Capital to improvements, which could be avoided have been the uncertainty which they have always found to attend the extent to which they would be called upon for their completion.[94]

The directors did not want to throw the blame onto the commissioners for the growing disenchantment with the company on the part of certain vocal settlers and reformers. Jones, particularly, had been having a difficult time in the Tract. After writing his *Defense of the Canada Company* Tiger Dunlop, who had recently been appointed general superintendent of the company, began to fall in increasingly with the Colborne Clique. Furthermore, the company steamboat continued to lose money and land sales were still dropping. The only solution seemed to be for Jones to spend more time in the Tract to order to put company business on a better footing.

The directors were sympathetic to Jones' problems and, noting his anxiety in correspondence, expressed "their sense of his zeal for their service on so many occasions."[95] They appreciated the difficulty of putting together reliable estimates and generally were satisfied with subsequent explanations when expenditures exceeded estimates. Still, the cost of undertakings invariably exceeded anticipated costs. Roads, the steamboat, the lumber account, the pier at Goderich and various minor undertakings for mills were all instances of this.

In their recommendations, Jones and Allan had suggested financial advances to the new District of Huron and for the London and Gore Railroad Company (which would ultimately become the Great Western Railway). In responding, the directors asked how the commissioners proposed to limit the advances to the amounts specified, and especially wondered about the estimates for the gaol and courthouse for the District.

> … what dependence can they place upon the sufficiency of the Estimate to complete the object, so that the funds appropriated

for repayment … shall not be still further charged to the prejudice of the Company?[96]

Allan and Jones had suggested financial support for the London and Gore
Railroad Company since the railway's promoters were planning to build a line
from a port on Lake Ontario to Detroit to divert U. S. trade through Canada,
rather than south of Lakes Erie and Ontario. As part of their grand plan for the
Canada Company, the commissioners proposed that the company build a line
from Goderich to Zorra Township to link into it. They argued that this line
would provide an all important shorter water route to and from points west.
Financial support to help make the Lake Ontario-Detroit and Goderich-Zorra
railway lines a reality would not only benefit Goderich in particular, but the
Huron Tract in general.

Looking ahead, the Canada Company directors were skeptical. More particularly, they envisaged the possibility of a link to Sarnia from the Lake Ontario-
Detroit line and the negative impact it would have on a Goderich-Zorra line.
They also worried about the cost of constructing such a line and its overall
operating costs. Furthermore, they questioned whether shareholders would be
interested in such a venture, noting that the company was in the business of
selling land, not building railways. Shareholders had already paid in upwards of
£200,000 and they were looking for a return on their investment from land
sales, including payment of arrears.[97]

Obviously, the directors did not want to engage in any operation unless all
was clearly defined. They did, however, give full credit to the commissioners'
arguments and their wish of "keeping up a good feeling in the Province."[98]
They were upset, though, at the treatment the company was receiving in Upper
Canada. There was no acknowledgment of the value of their marketing efforts
(particularly those made in Britain, France, Germany and the United States),
the improvements made on their lands nor of the information given to the
prospective emigrants by the company. They lamented:

> Can it be doubted for a moment that the Canada Company
> have been of very material use in this respect to the Province,
> and have indeed been the chief, if not the only channel of dif-
> fusing this information and with considerable success too, and

yet their efforts do not seem even to be admitted, or even adverted to in the Province. – So again regarding the assistance they have afforded to improvements – it must be remarked that they have never yet withdrawn one Shilling, but have constantly been increasing their advances, not only to the extent of the Capital raised, and transmitted there, but to the whole amount of all the profits realized upon their Sales, forming altogether a very considerable sum of outlay.[99]

The arrears were troubling the directors. They argued that "indulgence should have a limit."[100] From 1827 to 1833, sales had been made in Guelph to parties from whom the company had received no payment. Other purchasers had paid one-fifth down, or a part on account, and no more. Similar cases existed in the Huron Tract and on sales in the Crown Reserves and, because the interest was not being compounded, the company was earning less than five per cent on the arrears.[101]

The political unrest of 1837 and the economic difficulties which resulted were taking their toll. Sales of land were dwindling and instalment payments were falling off seriously. The company was forced to realign its financial structure given that the present state of affairs in Upper Canada would in all likelihood continue for some time. Its fixed costs amounted to £42,000 per year, to which expenditures in the colony had to be added.[102]

In authorizing expenditures in Canada, the directors had been guided by two considerations: that large and increasing overdue bills would be collected; and that the progress in the province in terms of wealth and population, along with increased expenditure sanctioned by the legislature for public improvement, would create a steady and improving market for the sale of company land.[103] These assumptions, if met, would have made the company self-sustaining thus negating the need for further calls on the shareholders, but this was not to happen.

The directors were upset with the estimates for 1836 and 1837 as no parameters for expenditures had been provided. In 1837, for example, cash expenditures were estimated to not exceed £500 as all outstanding accounts had been paid. Yet the outlay for that year totalled some £10,000, exclusive of salaries. Furthermore, over and above payments to the government for lands purchased, the commissioners had disbursed £80,000 from 1834 to November 1837.[104]

The directors recalled their experience with Galt some ten years earlier:

> You are aware that the Company have had some costly experi-
> ence in the amount of expenditure in improvements derived
> from the extravagant outlay upon the Guelph Settlement, the
> result will be that altho' no Serious check like the present was
> experienced, that after the disposal of all the property, a very
> inadequate profit will be derived from that quarter, and you
> will no doubt recollect how frequently the disappointment
> experienced there has been referred to as a caution against
> similar proceedings in the future.[105]

The company outlay on the steamboat from its inception to November 1837 had been £11,524.[106] It appeared to the directors that it had been of little use to the settlement, despite all of Jones' optimism. Its freight had consisted mainly of company lumber and only a few settlers had arrived on it. Furthermore, Captain Robert Dunlop had been relieved of his command for extravagance and incompetence. The boat had been a constant headache and a drain on finances.

Once again the directors called for a significant reduction in expenditures. All undertakings approved by the board but not yet begun were not to proceed, and all works in progress were to be slowed down, with completion deferred for as long as possible. Steps were also to be taken to reduce the size of the company's establishment in Canada and more pressure was to be brought to bear to collect overdue accounts. The commissioners were also to have a look at instituting a program of land sales by auction on former reserve lands. While neither commissioner favoured this approach, they were asked to assess this option because of the cost of maintaining company agencies in the eastern areas of Upper Canada.[107] The directors reasoned that if land could be sold by public auction, it would no longer be necessary to employ agents in this region.[108]

Despite the setbacks, the directors again expressed confidence in their commissioners in November 1837:

> they [the Court of Directors] give you full credit, as they have
> frequently expressed for [your] great zeal and anxiety to promote
> the interests of the company, and are willing to allow fully for
> the difficulties which have attended your proceedings.[109]

Site of W.L. Mackenzie's printing plant in Toronto. William Lyon Mackenzie was a Reformer. When Tory youth raided his printing plant and threw his type into Toronto Bay, the resulting trial gave him the publicity he needed for his radical newspaper. This building at Bay and Adelaide streets was on the site where plans were made which culminated in the Rebellion of 1837. *Courtesy of* The Globe and Mail.

The commissioners were going to need that confidence because events in December of that year and early in 1838 were to heighten company unpopularity among the Colbornites.[110]

When Toronto was attacked on December 7, 1837, by William Lyon Mackenzie and his band of rebels, rumour spread through the Huron Tract that the country was in arms and that invasion threatened from the south. Jones saw no reason to panic.[111] The Colbornites, however, did. Commanding officer of the First Huron Regiment, Tiger Dunlop, ordered blankets, food, a wagon and sleigh, and iron for pikes to be commandeered from the company. Jones countermanded the order, telling Dunlop to withdraw from his command. He refused to do so and, instead, resigned as general superintendent of the company the next month. Dunlop's friends lodged a protest with the directors for Jones' actions, as did Robert Dunlop as MPP. The reply from London expressed the directors' "unabated confidence in their confidential Officers"[112] and instructed Robert Dunlop to communicate through the medium of the commissioners in future.[113] In the meantime, the directors expressed their "entire approval" of Jones' acceptance of Dunlop's resignation,[114] and they requested the commissioners to have no further communication with either Robert Dunlop or Henry Hyndman on behalf of the officers of the Huron Militia over the affair:

> The Court are desirous to be explicit on this point in a full conviction that their Answers to the Several parties may justly entitle Mr. Jones to feel at perfect ease in respect to their excited and unfounded expressions of disapprobation.[115]

In 1838, with emigration seriously disrupted (only 3,239 settlers had landed that year in Quebec City) and gross company sales amounting to only £7,014/0/2,[116] the company became alarmed. The reduction in sales of land from a high of 114,804 acres in 1832 to a low 15,718 acres in 1838,[117] plus the reduced collection of accounts, including arrears, was having a very negative impact on company finances. With unsold land accumulating, collections falling and the company having to pay tax on land in their name, the government was approached to suspend the granting of additional land until conditions improved. This request was refused.

From its perspective, the general state of the company was approaching disastrous proportions because of the financial impact created by greatly declining land sales. A general "belt-tightening" was once more called for. Because of an ongoing money-losing lumber deal entered into by Thomas Mercer Jones in 1834, the directors decreed that "no such engagements" would again be agreed to without prior approval.[118]

In what appears to have been a means of keeping the company steamboat busy in the event that the hoped-for influx of settlers requiring transportation by water did not materialize, and to help the company's bottom line, Jones had developed a plan. To this end, he had entered into a contractual arrangement over a three year period to purchase all the lumber that could be produced by lumbermen Messrs. Brewster and Smart of the River Aux Sables region on Lake Huron at the southern end of the Tract. Jones' plan was to stockpile the lumber at Goderich and to sell it in Michigan whatever was not needed locally.

John Hicks and his father built a two-storey frame hotel in Mitchell in 1838, followed by the larger Commercial Hotel in 1857. Hicks was the first district councillor for Logan Township. *Courtesy of the Stratford-Perth Archives.*

Hicks House. Mitchell had a vibrant main street until it was wiped out by a fire in 1872. John Hicks built the Hicks House in 1873 on the site of his Commercial Hotel, which had burned. This photo is circa 1890–1900. The Davidson family purchased the hotel in 1882. It remained in their family until the 1940s. *Courtesy of the Frank Campbell Collection, Mitchell.*

District of Huron Gaol, Goderich. The Canada Company donated two-and-a-half
acres of land overlooking the Maitland River for the gaol. Opened in 1842,
it cost £4,868 to build. The stone for the two feet thick walls came from the river.
The gaol remained in operation until 1972 and is now a designated National
Historic Site. *Courtesy of the Huron Tourism Association.*

Presumably the lumber would be transported there by the company boat. The
arrangement was great for the lumbermen, but failed badly from the company
perspective. Not only was competitively priced lumber produced locally in the
vicinity of Goderich, but there was just as much timber in Michigan as there
was in Upper Canada. By December 1838 the company had taken a loss of
some £11,000 on the contract because of loss, rot and poor sales.[119] From the
beginning, the directors had doubted the wisdom of the undertaking and had
questioned Jones' on it. If the scheme were to be profitable, why could it not
be in the hands of individuals? If were not to make a profit, why go ahead with
it in the first place. Besides, they queried if it would not have negative impact
on the sawmill in Colborne.[120]

Although the directors had called for austerity measures, they were nonetheless committed to a loan of £3,100 for the proposed Huron District gaol and
courthouse, and they agreed to donate the sites for them. They decreed,

Court House, Goderich, 1858. In 1841, the Canada Company loaned the
District of Huron £3,100 for construction of a court house and gaol in Goderich.
The Court House actually cost £4,000 when completed and was considered
one of the finest in Upper Canada. It was destroyed by fire in 1954.
Courtesy of the John Graham Collection.

however, that the loan was not to be handed over in a lump sum, but rather
advances were to made as work proceeded. Furthermore, they said that on no
account would additional funds be granted should the project go over
budget.[121] In return for the donation of the site, they told the commissioners
that they expected that at least some of the funds advanced would come back
to the company through the employment of persons indebted to the company.
This, the directors reasoned, would save them from the necessity of undertaking any other works at Goderich.[122]

Given the financial state of the company, the directors continued to press for
more action on collections. Collections had been a touchy issue in view of the
unsettled political and economic state of the province. Furthermore, settlers had
been taking advantage of the commissioner's known intention of not enforcing

payment with the result that, in 1838, only £12,800 on a total of £100,000 owed in overdue accounts had been collected.[123] Capital was at a premium and as long as the company did not actively press for payment, many debtors were not prepared to come forward voluntarily. Moreover, the longer those in default were not pressed, the greater the difficulty in recovering the outstanding debt.

Because of the current unsatisfactory state of affairs and their view of Canada, the directors were very frank on the subject of public works:

> The Directors consider it one of the most important duties ... to take care that the improvements of the Company are all Works of absolute necessity, and conducted in the cheapest possible way to accomplish the objects, and without reference to appearance or magnificence of any kind. These are considerations which it will be time enough to entertain a century hence, and perhaps not even then. Canada is essentially the country for poor Men, and should be treated accordingly, and upon this principle only, can prosperity ensue at present.[124]

It was all very well for the directors to say this when safely in England, but the commissioners had to implement the company policy, and that presented many challenges. For instance, the commissioners had requested an additional pair of stones for the company gristmill on the London Road. The reply from the directors openly questioned the need for them. Before authorizing the purchase, they wanted to clearly understand the rationale for it. The latest return of the Huron Tract had shown that the number of settlers had actually decreased – "why should the mill need enlarging?" they queried.[125] The directors and commissioners were now beginning to work at cross purposes. The directors thought settlers should arrive in the region first, after which mills could be built by private individuals. The commissioners, on the other hand, wanted the facilities first in order to draw in the much needed settlers, along the lines of what Peter Robinson had done in his settlement of Peterborough.

The directors were becoming increasingly insistent that they wanted no part of running mills, steamships, stores or any business which required detailed superintendence. The business of the company was to sell land. Any departure from that would only be reluctantly undertaken where no individual could be found.[126]

In Retrospect

Considering the circumstances surrounding their appointment, and the conditions at the time, Thomas Mercer Jones and William Allan had been given adequate discretionary powers. It was only natural that they should have been held to a tight rein to begin with, following the uncontrollable Galt. The directors did not want to maintain control for control's sake, but they were treading a fine line – between the settlers and government on the one hand, and the shareholders on the other. Once authority from the London headquarters had been re-established by 1830, they were willing to loosen their hold as conditions warranted. In September of that year, for example, Allan had requested a relaxation of rules so that the commissioners, or whoever was to be in charge of Canada, could make the most of their situation at hand. He was thinking particularly of local projects which would benefit and promote the interests of the company, but for which there would not necessarily be sufficient time to obtain authorization from England. Based on their performance, this request was immediately granted, with a discretionary spending limit of £1,000 imposed where opportunity would be lost through reference to the directors.[127] Full control over the appointment of employees in Upper Canada, however, was not provided to the commissioners until 1836.[128]

The period from 1829 to 1839 had been at once one of adjustment, expansion and consolidation. The two commissioners had been responsible for putting the company in Canada on a business-like footing. They had both been the butt of criticism, particularly Jones, not so much because of themselves personally, but because they had to implement company policy which was not necessarily popular. Furthermore, Jones had been easier to criticize because of his role in travelling the company territory and spending up to six months a year working out of Goderich. He had more direct dealings with the settlers and, unlike his co-commissioner, did not have the status of being a member of the Executive Council.

The 1830s were challenging times in Upper Canada generally, both politically and economically. Communities were being founded and new political ideas formulated. Jones and Allan had made their mistakes during this period of uncertainty. Rarely were Jones' financial estimates close to the completed cost of projects. Furthermore, it would have been very difficult for the commissioners to divorce themselves completely from the pressing needs of the settlers.

The directors were generally aware of the difficulties confronting their commissioners and made a conscious effort to keep from being overly critical. When fault was found, it was usually tempered with a comment praising them for their zeal and hard work. Both men were loyal to the company and a mutual sense of trust between the commissioners on the one hand and the directors on the other prevailed, at least until the 1840s as it applied to Jones.[129]

William Allan had been appointed primarily because of his stature in the colony, but of course, his business acumen was a very important incentive. He had not been expected to devote all of his time to company matters – and didn't. That is not to say, however, that he did not pull his weight. In the first year, he undertook the extensive survey of company business procedures and contributed immeasurably to putting the company on a firmer footing by 1832 through his knowledge of local conditions, and as a banker. Once this had been achieved, he pulled back. In fact, that year he went on a six-month leave of absence, undoubtedly precipitated by a death in the family and his own serious illness which had left him very much depressed. He rallied but was becoming disenchanted with his Bank of Upper Canada board and the key members on it. Amongst other things, they would not support his attempt to sell the government's shares in the bank in order to give the bank more independence. He resigned the bank presidency in 1835.[130] This was fortuitous for the Canada Company directors because the Huron Tract was requiring more supervision. Allan was asked to devote more time to company affairs (with a raise in pay to compensate). He agreed to do so, at least in the short run, and Jones was now able to spend increasingly more time in the Tract.[131]

Given the times, and all things considered, Jones had done a competent job generally, despite errors in judgment over the lumber contract and the steamboat, and his less than stellar performance in preparing expenditure estimates. The directors had sided with him in his dispute with Dunlop and the Colborne Clique in 1837, and they must have been pleased with his marriage to Archdeacon Strachan's daughter, Elizabeth Mary in 1832, because of the strengthened connection with that centre of power – the Executive Council. Strachan was certainly a person to reckon with. He had ideas for his son-in-law too. In fact, he went so far as to suggest that it would be advantageous for all concerned if Jones were elected to the Legislative Council, as representative for the Huron Tract and one familiar with Canada Company business:

> ... it is usual & indeed necessary for each section of the Country
> to have some one connected with it in the Legislative Council
> & no man is more connected or indeed can be so much with
> this New Country as Mr. Jones.[132]

The directors did not approve this suggestion. They did not want their people involved in politics, with the exception of William Allan who as a highly esteemed member of the Executive Council could look after the company's political dimension there. Having Jones on the Legislative Council would have put him directly in the middle of simmering political issues in Upper Canada in general, and in the Huron Tract in particular, which was the last thing the directors would have wanted. Jones was to devote his full time to the company – period!

Jones was also approached by the Bank of British North America to become a director of their Toronto branch. To this end, he happily apprised the directors that he had agreed to this opportunity. They expressed surprise at this turn of events:

> they think it right to observe that knowing how much Mr.
> Jones' time is occupied in the business of the Company, and
> that the still increasing extent of their affairs in the Huron Tract
> must create an increased demand upon it (where his occasional
> or entire residence would be very desirable) and with which
> such an engagement might interfere. – They hope that this is an
> unfounded report.[133]

The reason for Frederick Widder's appointment as commissioner in the spring of 1839 is not entirely self-evident. The official reason advanced by the directors was that they wished to find a third commissioner upon whom they could depend "in case of such misfortune as Mr. Jones' illness or inability from any circumstance, to carry on the details of the business ..."[134] Perhaps Jones was getting too involved in local politics and they wanted to ease him out. Furthermore, as a commissioner who was John Strachan's son-in-law and who when not travelling the Huron Tract territory was living in grand style in York, they may have thought of him as an easy target for the Reformers. A third commissioner would be an important "safety valve" for the company. If Jones were

Drawing room of Frederick Widder's home, "Lyndhurst," in Toronto.
The Widder family lived well in Lyndhurst on Front Street (east of Old Fort York).
Frederick and his wife, Elizabeth, entertained in style. *Courtesy of the Toronto Public Library*
(TRL), J. Ross Robertson Collection, JRR 884.

to leave, or be asked to leave for any reason, and/or if the directors were deciding that Allan should go, they would have a person being groomed to take over
from one or other, or both of them.

On the other hand, the company may have been using Jones as the excuse
to hire Widder, but in fact, wanted Allan to go. Past his prime at seventy, Allan
still a force to reckon with in Upper Canada, but was his Executive Council mem-
bership becoming a liability? The company was not getting much support from
government and had been coming in for criticism from all sides. It was time for
change – was the company too close to the government for its own good?

Although not a known quantity in Canada at the time, Frederick Widder,
who was born in England in 1801, was the son of Charles Ignatius Widder, a
London-based director of the company. He not only had the right connections
at the London end, but he and his wife were connected to royalty (he, Austro-

Bavarian, and she, English).[135] Widder was highly regarded as "an amiable, hospitable, organized, business-like man of moderate of temperament with solid administrative talents"[136] – just the right combination for the job ahead. Upon arrival in Canada, he became active in the St. George's Society and as lay vice-president of the Anglican Diocesan Church Society, concerned himself with the church's endowments. While these pursuits ensured that he was well-connected in the colony, he was not part of the government structure, but that would not now be a disadvantage for the company, particularly since it would be two years before William Allan would retire. In fact, being independent of government could be seen as an asset. He and his wife Elizabeth and their two children Blanche and Jane (and two others, a son and a daughter until they died in 1849) would live a life of style and opulence in their home, "Lyndhurst," overlooking Lake Ontario east of Old Fort York. Widder had enlarged the original home there quite considerably and Lyndhurst was to become "the centre of social attraction."[137]

CHAPTER 5

A STORMY FOURTEEN YEARS

Nusquam Meta Mihi – "Nothing Daunts Me"
(Widder Family Motto)[1]

THE PERIOD FROM 1839 TO 1853 was both a time of growth and a coming of age for the company and for the country with the union of Upper and Lower Canada in 1841. The Rebellion of 1837 was over and, in 1839, Lord Durham[2] presented his historic report which, in part, condemned the Family Compact in Upper Canada as "a petty corrupt insolent Tory clique." He called for responsible government with the executive being drawn from the majority party in the Assembly, and recommended the union of the Canadas. Following the union of 1841, responsible government was implemented over the period of 1847-49. Exports recovered, but Canadian wheat sales were still depressed until the prosperous 1850s which followed the repeal of the British Corn Laws in 1846 and the loss of preferential tariffs. Equally importantly, emigration to Canada reached and surpassed the pre-rebellion and pre-cholera level of 1832. By the time Thomas Mercer Jones was dismissed in 1852, the company was in its strongest financial position ever.

One must not be left with the impression, however, that it was an era of uneventful growth for the company. On the contrary, in 1840, for example, the company had to face a Commission of Enquiry while simultaneously dealing with an illegal tax that the Huron District Council had imposed on the company lands in the Huron Tract.[3] The resultant dispute would last for six years. Furthermore, there would be attacks by John Strachan under the pseudonym of "Aliquis," and attacks by W.H. Smith in his publication *Canada: Past, Present and Future*. This was also the period of the establishment, operation and then closing of the Goderich office for the Huron Tract lands and Jones' tenure there on a full-time basis. The directors had to contend with Thomas Mercer Jones in 1852 over the railway issue – he was either being "bloody-minded" or naive. The period would be marked by William Allan's departure as commissioner, by the dominance of the Colbornites on the Huron District Council, and by Jones being "taken in" to support interests opposed to those of the directors. The election of 1841 was controversial and created adverse publicity for the company, while Widder's celebrated leasing plan (not unlike one proposed by J.B. Robinson some nineteen years earlier) gave the company very creditable publicity in the Canadas and Great Britain. Throughout the period, a growing animosity emerged between Jones and Widder, each of whom attempted the implementation of schemes to improve company land sales, impress the directors and "best" one another. Widder emerged the undoubted winner.

Until Widder's appointment, Jones, in reality, had been the senior company commissioner, although technically speaking Thomas Mercer Jones and William Allan were equals. Allan had always stayed in Toronto and had devoted only a portion of his time to the company, while Jones had devoted his full time. With Widder's appointment, Jones would unknowingly have a competitive rival whose executive talents would certainly surpass his, a fact which the directors may very well have realized.

The Commission of Enquiry

The establishment of the company's Goderich office in 1839 and the appointment of the colonial legislature's Commission of Inquiry in 1840 to look into the affairs of the company are not directly related, although the seeds of both were sown almost simultaneously as far back as 1830. In the final analysis, it was in spite of, and not because of, the establishment of the office that the Inquiry was launched. The Colbornites had been simmering for some time, especially after the episode with Dr. Dunlop in the rebellion period of 1837-38 which had led to his resignation from the company's service. Furthermore, Jones'

Canada Company Headquarters, Goderich. Built in 1839, the Canada Company
headquarters for the Huron Tract also provided living accommodation for the
Jones family. They loved to entertain and had many lively parties here. It is now the
Park House Restaurant and Pub. Company superintendent John Longworth
oversaw its construction. *Courtesy of the Stratford Beacon Herald Archives.*

behaviour in 1839 during a visit to Goderich by the new lieutenant governor,
Sir George Arthur, only heightened their distain for company practices and
caused them to press with even more vigour for an official inquiry into the
affairs of the company.

There were good reasons for invoking William Allan's proposal of nine years
earlier to establish a permanent office at Goderich in place of the agency which
had been operating under Dunlop's charge. Not only was it the best way to
maintain tighter management control in the Huron Tract, but it was to the
company's advantage to effect sales as quickly as possible in order to avoid paying
tax on unsold land. A permanent company commissioner in the Huron Tract
would facilitate this. It would also add prestige and speed up company transactions. Thomas Mercer Jones was judged the obvious person to send to Goderich.

Appointed to the position in 1839, he moved to Goderich with his family
in 1840 along with twenty-one wagon loads of goods and chattels. A new
office/residence at the top of the harbour hill had been built at a cost of

£2,222/3/0.[4] His twelve-year tenure as resident Canada Company commissioner there was about to begin.[5]

It must have been with a sense of relief that Jones received the Goderich assignment, as in that first year after Frederick Widder's appointment there was no clear indication as to whether the directors were planning to keep three commissioners, or to release one of them – which would mean either Allan or Jones. He was apprehensive and wrote to the directors, asking for an explanation as to why no specific duties had been assigned to each of the three commissioners. He was only satisfied when the directors replied that they were expressing their confidence in the discretion and judgment of the commissioners by not doing so. It was hoped, they said, that this explanation would remove from Jones' mind the uncomfortable impression he appeared to have received.[6]

Apparently Jones was still well thought of by the directors despite his sometimes improvident transactions. The visit by the new lieutenant governor had given him the opportunity to show off the town founded by the company on the shores of Lake Huron, and the directors had expressed their complete satisfaction in their commissioner, having received his report on Arthur's visit:

> It is very gratifying ... to learn, that the Lieut. Governor Sir George Arthur has visited Goderich, and that he has spoken so favorably of the progress the Company have made in settling the Huron Tract, and also of the different undertakings they have completed, or are in progress of completion there. The Court feel that the satisfaction thus expressed may be very much attributed to Mr. Jones' indefatigable exertions.[7]

Jones had neglected to tell the directors the devious means he had used to impress Arthur. The rationale for the official visit was to assure the lieutenant governor that the works undertaken by the company, and charged against the payments made to the government, were being properly carried out. During the course of the visit, not only did Jones annoy the Colbornites, but he also snubbed former company employee Captain Robert Dunlop, who was now a member of the provincial parliament (MPP) – and a Colbornite, as it will be recalled. Furthermore, Jones also used devious means to impress the vice-regal party as to the intensity of the work being carried out by the company.

His first *faux pas* was when he placed company engineer John Longworth in

John Longworth as an older man. Thomas Mercer Jones replaced Charles Prior, his disgraced Canada Company superintendent, with this man who, in earlier years had walked out on his wife and seven children in Ireland. Longworth was not liked by the settlers, and his rough and domineering character did not bode well for the company, but Jones liked him. In later life, Longworth did become a respected member of the community. *Courtesy of the Stratford-Perth Archives.*

the place of honour beside Sir George Arthur in the official party instead of Robert Dunlop. Furthermore, just prior to the visit to the harbour, every available man in the community was mustered to work on the company pier. After the harbour inspection, the vice-regal party was then taken on a circuitous route to the site of the Colborne bridge project, allowing time for the men at the harbour to rush over to the bridge project. The lieutenant governor did not know that he had been seeing the same workmen twice. To conclude the day, Jones held a garden party, a dinner party and a ball.[8] Arthur was very suitably impressed, but to the Colbornites, Jones' behaviour was classic. From their perspective, the company had duped the lieutenant governor for its purposes and had not given due recognition to the local MPP – not a smart move on Jones' part.

By this time, the Colbornites, along with the reformers, were determined to press for a Commission of Inquiry to investigate the affairs of the company. As well, Messrs Lizars, Hyndman and Dunlop were all writing to either the *Toronto Patriot* or the *British Colonist* and expressing their indignation at the company. Jones replied through the newspapers in March 1840:

> … the more the official conduct of the Canada Company, & of
> their commissioners in this country, is enquired into – provided

only the result of these enquiries be faithfully communicated, the higher will be the situation of the Company in the estimation of the public.[9]

This opinion was echoed by the directors in December of that year. They reiterated that no "evil results" would arise to the company's interests from the inquiry. The explanations of the commissioners, they said in a letter to Jones and Widder, if candidly considered, "will show that the company has faithfully fulfilled the engagements under their Charter, and has greatly promoted the interest of the Province."[10] The directors informed the commissioners that should there be a formal inquiry, complete information was to be provided. However, no amendment to the charter was to be tolerated, nor was any attempt to impose new conditions on the company to be sanctioned.

As far as the directors were concerned, how the company paid for improvements was not open to discussion. Whether the contractors were to be paid in money, or in land as part payment "cannot be a question of which the public have any interest, or right of interference."[11] The contractors, of course, had little choice in the matter. They either had to agree to accept land as partial payment or lose the contract – a reality that the directors were not prepared to admit. Realizing the implications of a single contract, the commissioners were directed henceforth to negotiate two contracts. One would stipulate a monetary payment for work done, while the other would stipulate payment in the form of land at a stated price per acre. The directors advised the commissioners to use this formula as much as possible in order to speed up land sales, and to limit the outlay of "monied capital."[12]

The pressure applied by the Colbornites, and by their member of the provincial parliament, was sufficient to force the government to create the Commission of Enquiry in 1840, and a commission of three was duly appointed.[13] However, the exercise came to naught because any criticism of the company over expenditures in the Huron Tract, taken out of the Huron Tract Improvement Fund, would fall directly on the Executive Council since technically the company was obliged to submit, and indeed did submit, all estimates to the council before undertaking any work. That said, some evidence was collected. A preliminary and confidential report of one of the commissioners, Mr. Justice Jonas T.W. Jones[14] of the Court of King's Bench, dated November 2, 1840, indicated the path which the inquiry was taking when disbanded. The report indicated that the directors were not far wrong in their assessment of the commission's findings, although for different reasons.

A visit to England on private business had prevented Jonas Jones from completing his inquiry. He began his report by noting that the inquiry commissioners had thus far only obtained information from persons hostile to the company. Thus, the report was less favourable to the company than would otherwise have been the case. The essence of his findings was that while some expenses had been wrongly charged to the Improvement Fund, the company could not be called upon to refund the money which had been placed to its credit. In each case, approval of the Governor-in-Council had been obtained as required. Jones did criticize the cost of both the Maitland River bridge (£3,731/16/5) and the approach road to it through Goderich (£405).[15] Furthermore, he did not agree with the company practice of paying for public works partly in cash and partly in land because of the resulting higher cost, which would be charged against the Fund. He also noted that as an inducement prior to the signing of the final agreement with the government, the company had held out the prospect of spending upwards of £20,000 yearly on improvements, in addition to the annual payments to the government. This had not happened:

> if the half of this sum had been expended annually in improving the Company's lands the present state of their settlements would be very different from what it is.[16]

While he criticized the lack of mills, Jonas Jones stated that a detailed report by Daniel Lizars which roundly criticized the company contained many points which were not quite correct. He also noted that the spirit in which the Lizars' report had been written was very obvious.[17]

Mr. Jones summed up his report by concluding that nothing further would have been achieved if the inquiry had run its natural course. The company had been given extensive powers, and because of the way the articles of the agreement had been drafted, it would have been shielded in most cases. On almost every point, the company could have said it was within the letter of the law and:

> Even if the whole amount objected to in the reserved funds could now be made a proper charge against the Company, it would not be of any great public benefit, as sooner or later it is

obvious that the Company must for their own interests expend a much larger sum in public works.[18]

Reassessment of Company Policy

If Jonas Jones actually thought the company would increase spending on public works, he was off the mark. At precisely the same time as he was preparing his report, the directors in England were reformulating policy as it applied to company expenditures in Upper Canada, including expenditures on public works and company personnel. As part of the process, William Allan would be terminated and Frederick Widder would begin his ascent to the unofficial position of senior commissioner in Canada.

In a lengthy letter to the company commissioners in October 1840, Frederick Widder and Thomas Mercer Jones were told to cut expenditures immediately. The directors noted that up to December 31, 1839, a total of 197,612 acres had been sold in the Huron Tract, producing £111,253/17/9, while company costs with interest at six per cent amounted to £111,540/16/11, not including the capital invested in the Goderich pier and mills, and advances on mortgages.[19] The directors were convinced that unless sales could be accelerated considerably, or higher prices obtained for land (a doubtful prospect at this stage), nothing but a large-scale reduction in the expenditures could save the company from operating at a loss. It was for this reason, they said, that they could not justify retaining the services of three commissioners:

> … as two were found sufficient when the business to be performed was as difficult, if not more so, than it is at present, or likely to be in future: On this point therefore, the Court propose to address a letter to Mr. Allan by this opportunity.[20]

Now in his seventieth year, William Allan had already decided to divest himself of some of his responsibilities, including his positions as commissioner for the Canada Company and as member of the Executive and Legislative councils. Accordingly he submitted a letter of resignation to company directors in the autumn of 1840 which, by coincidence, crossed in the mail with the company's letter of termination. He would now be in a position to mange his very extensive land holdings throughout Upper Canada and other interests there.[21] He was a shareholder in the Bank of British North America which was founded in 1836 and he had become the governor of the British America Fire and Full Life Assurance Company that same year (and served in that capacity until his death in 1853). He was also voted president of the City of Toronto and Lake Huron Rail Road Company in 1837 (but because of the unsettled times, the railway project was not revived until the mid-1840s). His work as a trustee of major estates also occupied much of his time as did his handling of former Canada Company director Edward Ellice's land speculation and financial matters in Upper Canada. As if that were not enough, he chaired meetings of the British Constitutional Society and, at age 79, became a vice-president of the British American League. He also continued in his support of the Anglican Church by serving as a board member of the Church Society and handling of the funds for the Anglican Diocese of Toronto. Last but certainly not least, he was a strong supporter of Bishop Strachan in his efforts to found Trinity College and was on the board of Trinity when he died at age 83 (some say from "sheer exhaustion).[22]

In his letter to Mr. Allan, advising that the services of this high profile Canada Company commissioner could no longer be justified, company chairman Charles Franks stated that it was unnecessary to comment further on the measures of retrenchment which had become unavoidable, but that the directors were "sensible of the advantage" the company had derived from his "great experience," more especially in the early stages. Franks concluded by noting that as the directors had not expected Allan to devote his full time to the company, they wished to employ a third commissioner to furnish those details promptly and "to establish an active correspondence." Because business had not increased as they had hoped, a reduction in the expenses of the establishment had became necessary, he added.[23]

The directors once again developed a hard line on company expenditures. No further outlay for improvements was to be undertaken beyond those already authorized and those already sanctioned were to be done gradually. Future improvements in the Huron Tract were to depend on the inhabitants. A reduction in the price of the lands for sale was contemplated and the reintroduction of an allowance for travelling expenses from Quebec was discussed. Expenses for road maintenance were now being considered because the directors realized that settlers were unwilling to effect repairs on the roads, even in sections where the company had sold most of its land. Specifically, they were concerned about

The Huron Road circa 1858. Cut through by Tiger Dunlop in 1827, the Huron
Road from Wilmot Township to Goderich was widened from a primitive trail,
and then a "sleigh road," to 66 feet in 1829. The original road was surveyed by
John McDonald, Deputy Provincial Surveyor. Burnt stumps still remain in the field.
The board fence indicates that the owner was of a certain wealth.
Courtesy of the Archives of Ontario, P 129 ACC 1327 AO 6476.

the road from Goderich to Bayfield because Thomas Mercer Jones had estimated that £500 would be required to put the twelve-mile stretch back into
passable condition:

> it is too obvious that if the Company undertake such expendi-
> tures, there will be no limit to the expectation of the Settlers,
> or to their dependence upon the Capital of the Company.[24]

Reluctantly, the directors agreed that if the company was compelled to do the
necessary remedial work, the government-legislated statute labour due by settlers would be strictly applied.[25] The directors were no longer prepared to absorb
the full cost of improvements. The settlers would have to do their part too.

With the arrival of Frederick Widder, the company would once again go
through a period of reassessment, including the development of a more sophis-
ticated management structure. They not only had to get their finances under
control, but they had to encourage more emigrants to settle on Canada

Company land. To this end, they discussed re-instituting the free conveyance of settlers from Quebec to the Huron Tract, a service that had been terminated some time earlier because of its being underutilized. In fact, the company petitioned the Colonial Office in London for permission to apply the cost of doing so, plus the cost of promoting emigration to Canada, against the balance presently due the Crown (£60,000).[26] While the British government appeared sympathetic to the company's plight, the request was turned down. They replied that company payments still formed part of the revenue of Upper Canada and while the government offered to place the funds at the disposal of the legislature of Upper Canada in exchange for the Crown paying government employee salaries, that offer was still before the House of Assembly. Accordingly, the proposal could not be entertained.

While the proposal apparently came to naught, the British government did make a concerted effort in the 1840s to increase emigration to Canada through the efforts of the Colonial Land and Emigration Commission in 1840 and a multitude of societies in England, Scotland and Ireland who were actively engaged in promoting emigration in order to ease the suffering of the poor. These efforts in turn were helpful to the company. In the meantime, however, Thomas Mercer Jones, having been criticized on a number of occasions for spending funds unwisely, was determined not to get caught again and so advised against reintroducing free or assisted passage from Quebec to Hamilton, and implementing steamboat service from Fort Erie to Goderich. The directors were perplexed, but gave him the benefit of the doubt:

> The Court regret that you have altered your determination to encourage Settlers in the Huron Tract by payment of their Travelling expenses to Hamilton. The Court consider it a very judicious step at this moment, and should have thought it wise to extend it to persons coming from the United States by the Steamboat …. The Court however desire me to say that this is a matter which might be so much better understood on the spot than it can be in London.[27]

Jones subsequently proposed an alternate plan for steamboat service from Detroit to Goderich and a stage line from London to Goderich. It was accepted and implemented.

With renewed efforts by the company to increase settlement on its lands, it was vital that both the directors and commissioners have accurate and well-authenticated reports on the results of emigration and settlement on Canada Company lands. Accordingly, Widder and Jones were directed to draw up annual reports as follows:

> 1) a statement of the actual circumstances of all those persons purchasing company lands, including the date of each purchase and when settled;
> (2) the amount of capital possessed, if any, by these persons;
> (3) a description of the family, including number;
> (4) the place of origin of the settler/family;
> (5) a description of the settlers' present situation, i.e., quantity of cleared land, whether paid for, type of house or other building, number of cattle and description of any other property.

The first return was requested for the year 1840.[28]

Frederick Widder's Increasing Prestige

Frederick Widder was quick to seize the opportunity to provide a detailed annual report. He did not have specific responsibility of the Huron Tract and, although company Crown Reserve sales were twice those of the Tract, there were fewer issues to deal with concerning these properties so he had time to develop, influence and implement company policy more so than did Jones.[29] Like his predecessor John Galt, however, Thomas Mercer Jones was becoming too involved in local issues, contrary to company policy. This was not only time-consuming, but proved embarrassing to the company. The election of 1841 was a classic example and because Jones attempted to satisfy everyone, he was never really able to completely satisfy anyone. Widder was much better at public relations than Jones and remained above the fray. He avoided contentious issues and concentrated on burnishing his image vis-à-vis the directors while at the same time improving the lot of the company. His most significant achievement in this regard was the leasing plan.

For his first year in the company, Frederick Widder had not openly disagreed with Jones. He had generally co-operated with him, while studying agricultural and economic trends in the colony and focusing on company policy issues. Once he felt himself established, however, he began a voluminous correspondence with the directors and often differed with Jones on matters of policy. A co-signed letter from Widder and Allan to the directors is a case in point. These commissioners did not agree with Jones' negative opinion on the plan to re-introduce free passage from Quebec City for company settlers and noted:

> We regret … that his [Jones'] opinion is decidedly adverse in every way, to the principles & expediency of this measure, … yet having reference to his great experience & knowledge of the Huron Tract, & the consequences he strongly apprehends, we saw no alternative left to us.[30]

On the assumption that Jones would agree to the free passage proposal, Allan and Widder had actually made all the arrangements for the conveying of prospective settlers to Hamilton from Quebec City. These plans then had to be cancelled because of Jones' intransigence, and the commissioners were now limited to printing and sending a list of company lands for sale to the company agent in Quebec City and the government agent in Montreal. This list was then subsequently distributed "at every place on the route from Quebec to this [Toronto]." They also had a list of lands in each township printed and made public "in the most desirable manner in the respective Townships."[31]

From this juncture, until Jones' dismissal in 1852, a growing rivalry developed between Jones and Widder, with Jones always seeming to come out second best. Widder was more astute than Jones and was able to sense the mood of the directors and propose policies likely to benefit the company and please the directors, whether relating to advertising, sales or public relations. Although Widder's suggestions were not always accepted, particularly when it came to company involvement in the establishment of a business, such as his proposal to establish the "Lake Huron Fishery" which would have involved the company in fish trade, the directors were obviously satisfied with his work. In February 1841, they increased his salary by £200 sterling to £600 sterling per annum which was the same salary Allan had been drawing when he retired.[32]

While Frederick Widder involved himself in developing ideas to increase company sales, the not-soon-to-be-forgotten election of 1841, following the union of Upper and Lower Canada that year, was taking much of Jones' time and effort in the Huron Tract. Unfortunately for the company, and primarily because of Jones, the election in the Huron Tract became one of the Canada Company versus the reformers. Robert Dunlop, member of the provincial legislature from the Huron District, had died in February of that year and his brother, Tiger, had been prevailed upon to run in his place. Instead of attempting to pursue a middle course, as preferred by the directors of the company, Jones chose to become involved by nominating company engineer John Longworth to oppose Dunlop. The nomination of Longworth created such a furor locally that he was forced to withdraw. Jones then proposed his brother-in-law and company legal counsel in Goderich, Captain James Strachan, in place of Longworth. The election was hotly contested and after a week of voting, Strachan was declared winner by ten votes. Dunlop's supporters cried "foul" and petitioned for an inquiry into what they viewed as election irregularities. They alleged that many Strachan voters should have been disqualified as they were underage and/or did not own land or if they did, had not owned it for the required twelve months. A legislative committee was duly appointed to look into the matter; it determined that fifty-nine votes for Strachan should not have been allowed. Dr. William Tiger Dunlop thus became the first member of parliament from the Huron District in the legislature of the united Canadas.[33]

> The directors did not want to believe newspaper reports of his involvement: Mr. Jones may feel assured the Court have such confidence in his caution and zeal for the Company's interests, that they are not likely to be led away by any statements, particularly those made in Newspapers, until they have received his own explanation.[34]

The directors reminded Jones that as a representative of the company he should not have meddled in politics. They had not seen the speeches of the candidates, but said they could not imagine that opposition to, or support of, the company's interests were grounds for appealing the outcome of the vote. The interests of the company were, or ought to be, "so identical with those of the province," and so completely a part of those of every community for which representatives were chosen, that such an appeal in the eyes of the community should appear "groundless and absurd." This, of course, was far from the case, for it was political "hay" to attack the company for all the perceived wrongs in the community, whether legitimate or otherwise. The problem was, Jones did not have the sense to remain impartial as the directors had decreed to their officers and staff in Canada. Despite Jones' unwise involvement in the disputed election, his explanation after the fact must have satisfied the directors as there was no further mention of it in later correspondence.

In the meantime, while Jones continued to get more deeply involved in local Huron District issues, Widder was working on the overdue payments issue. He wrote to all settlers in this regard and took selective legal proceedings against settlers as a warning to other defaulters. He was congratulated for his work by the directors at the same time as Jones was being criticized for his election activities.[35]

In what would appear to be an attempt to vindicate himself, Jones had prepared a report for the Executive Council in August 1840, in which he claimed a deficiency of 53,237 acres in the Huron Tract land. In doing so, he requested an equal amount of land north of the Tract. Jones was caught off-guard when the reply came back, for in fact, said his interlocutors, the company had been given an extra 23,849 acres. The committee of the Executive Council which was looking into the matter said, however, that they were willing to let the matter rest as they considered some of the land granted to the company to be inferior, that the company had to suffer any loss resulting from inaccurate surveys and that public roads, although included in the amount purchased, had remained the property of the Crown. They, therefore, suggested that the company pay for the 1,000,000 acres as originally agreed, "thus bringing to a simple and satisfactory conclusion all matters between the Canada Company and the Government relating to the Huron Tract."[36] Jones reported the outcome to the directors, who reluctantly agreed to accept the committee report after having consulted their legal advisors in England.[37] Jones should probably have known better than to involve the Executive Council in this delicate matter at this point, given the tensions of the day in the province with regard to the company. While the issue turned out to be a non-issue in the end, it did cause the directors some grief in the meantime and called into question Jones' political acumen which was in sharp contrast to that of Widder.

Widder, however, did not completely escape criticism. For reasons puzzling to the directors, the commissioners were at odds with London over expenditures in the Huron Tract. Although the directors were insistent on strict economy, their constant reminders on this issue implied that the commissioners were not paying attention to the dictates from headquarters. Of late, the directors had stated categorically that the financial outlay in the Tract had been decidedly too large for expenditures on, for example, roads, bridges and the Goderich Harbour, especially when considering the price for land at this stage.

In order to bring the financial position of the company to their attention more forcefully, the directors tried a different tack and requested the commissioners to file a detailed report as follows: (a) the average number of acres expected to be sold annually; (b) the estimated average selling price per acre; (c) installment payments which could be realized. The average annual expenses "of the establishment," including the taxes on the wild land and the annual outlay for improvements, was to then be deducted. Finally, the monies to be realized by installments were to be compared with funds already spent, including the amount raised by way of debentures, to which was to be added compound interest.[38]

While the ensuing report has not been located, suffice to say that the request did serve as a focus for the commissioners – and directors – as to expectations vis-à-vis future income versus expenditures.

The directors were not prone to criticize the commissioners unnecessarily, but they certainly did so when warranted, and it is most evident that Thomas Mercer Jones was the target for most of it. Furthermore, Jones did not appear to have the strength of character to deal with his domineering father-in-law, nor to fend off the endless demands of the settlers for an improved harbour, and new or improved roads, mills and bridges. He also found it extremely difficult to be answerable to the directors in London and to what he thought to be his duty towards the District Council in the Huron Tract. His attempt at compromise only resulted in general dissatisfaction all around.

The much needed bridge over the Bayfield River at Bayfield, twelve miles south of Goderich, was a case in point. Following Baron de Tuyll's departure from Upper Canada to his native Belgium in 1835 and his death that year, Vincent Gildemeester de Tuyll, the Baron's son, took over his father's properties. The wily second Baron, who had had little choice but to build a gristmill at Bayfield in the late 1830s, convinced Jones that the most economical method of getting from one side of the river the other was by way of a road on top of the mill dam (the Baron was to pay for it, with the company providing a loan to pay for its construction). Upon learning of this plan, the directors queried Jones on the efficacy of it. What assurance, they asked, could be given that the dam would be maintained or that it was sufficiently strong to support a road? No sooner had

The first Bayfield Bridge and the "New Bridge," Bayfield (circa 1910).
Bridges were always an issue. Who should pay – the company or the settlers?
In 1840, Thomas Mercer Jones authorized the construction of a road atop the mill
dam at Bayfield without the directors' sanction. The dam burst the next spring – the
directors were not pleased as they had predicted as much when they heard about it.
The company built the first bridge in 1841. *Courtesy of the Bayfield Archives.*

they questioned Jones' judgment than the dam washed away in the spring flood
in April 1841.[39] Jones would not openly admit the folly of his decision and it
was not until December 1841, when he referred to the need of "replacing the
Bridge," that they were able to conclude that the dam had in fact been washed
out.[40] "This is precisely the result apprehended by the Directors" wrote Perry,
"and referred to, when writing on the subject of your arrangements with the
Baron de Tuyll." He added "you ought not to be surprised at the expression of
the dissatisfaction of the Court at the general result."[41]

Consciously or unconsciously, Jones tended to skirt issues and to mislead
the directors as to actual circumstances in the Tract. In January 1839, for
instance, he reported that the mill properties owned by the company were fully
worth their actual cost value, "with no reservations."[42] Yet, when the mill at
Stratford was sold later in 1842, the company sustained a £400 loss.[43] An
example of his lack of thoroughness and detail were his weekly returns from
Goderich. The returns, complained the directors, contain so few details as to
not make them worth the expense of postage.[44]

Jones was repeatedly instructed to limit expenditures in the Tract to the
amount remaining unappropriated in the Improvement Fund and, in any case, to

notify the directors before any contemplated expenditure was openly discussed with the authorities or interested persons in the Tract. Yet he still sent requests (to the directors) for additional funds . In July 1842, for example, he asked for permission to spend £2,000 to equip two schooners for William Geary, a resident of Goderich who served on the Huron District Council, so that he could engage in trade with Goderich. He also requested the sum of £5,000 in order to establish a shipping company.[45] Company secretary Perry remarked of these requests, "the Court think he would have acted more judiciously if he had declined to send these proposals home for consideration."[46] He opined that if the business of the settlement held out sufficient expectation of a profit, individuals should be found with capital to support it, and not the company:

> The Court had really hoped that from the full and confidential manner in which their correspondence with you had been conducted for many years, that you would fully understand the situation of the Company, and were so entirely possessed with the views and opinions of the Directors that they might depend upon you for a corresponding action; they are however led to think ... that he at least misapprehends the position and wishes of the Company and seems to consider that some undefined fund exists in London to which recourse may always be made by the Directors for money.[47]

The directors went on to remind Jones that almost £30,000 had been paid in by the shareholders to complete the purchase of lands, and another £90,000 had been borrowed for debentures.[48] The proprietors did not look for an early repayment of their capital, they said, but interest payments alone amounted to £22,500 a year, and the debenture issue would have to be provided for in due course.[49] Furthermore, they noted, payments for land were not accruing as expected, with arrears to December 31, 1841 totalling £13,000.

It was not long before Jones was taken to task again. It had resulted from a request to increase his staff in Goderich by one in order to cope with increased business. He must have been surprised and frustrated by the response. Instead of granting the request, he was told that the company books were to be consolidated in Toronto, that company office hours were to be extended and that his staff complement was to be reduced by three. He was to dispense with the

First St. George's Church, Goderich. This Anglican Church was built on St. George's Crescent in 1843 on land donated by the company. Until this time, parishioners worshipped firstly in a log schoolhouse on East Street and then in a "stable church" on West Street. The church burned in 1878 and was replaced by the current St. George's in 1881.

Courtesy of St. George's Anglican Church.

services of company engineer John Longworth (because no more works were contemplated), along with the bookkeeper and accountant. Henceforth, all entries were to be recorded by means of a cash book and journal, with copies sent daily, or as frequently as possible, to Toronto. These would then be incorporated into one set of books in the company's general accounts. The directors told Jones that they had decided to implement this procedure because the books for the Huron Tract had proved to be such a source of trouble and expense, "as well as a great inconvenience" due to numerous errors:

> … nothing has occasioned so much difficulty at home, as the necessity of reconciling and correcting the very numerous errors, which have arisen from your Bookkeeping.[50]

Despite having been chastised five months earlier for laxness in collecting installments, Jones had still not learned:

> Hitherto the plans pursued in the Huron Tract seem to proceed upon no regular system, sometimes Settlers are required to pay a first instalment and sometimes not, without any apparent reason, and as to subsequent instalments there seems to be no

> regularity whatever, or any attempts at enforcing any rules, and it appears to the Directors that the Settlers in the Huron Tract rely so entirely upon the indulgence that has been afforded them, that they think it altogether unnecessary to give attention to their obligations.[51]

Jones did not take kindly to this criticism and wrote the directors expressing his dismay. They responded: "those observations were intended as instructions for the future and not as condemnation of the past."[52]

Although the directors continued to consult Jones, Widder was increasingly being looked to for his opinion on matters pertaining to the Huron Tract. The perennial expenses at the harbour, due primarily to storm and ice damage, resulted in a *joint* report being requested as to the future of the port. The directors were becoming skeptical of Jones because of his unreliable estimates of expenditures. Not only did they invariably prove to be wrong, but Jones was prone to exceed imposed spending limits. He had been authorized, for instance, to advance £2,500 to Surveyor John McDonald for the new mill at Goderich, on the security of the mill, but then went ahead and loaned him £3,000. Company secretary Perry asked for an explanation pure and simple: "The Court beg to know why you have so much exceeded it."[53] The directors had not wished to become involved in the mill in the first place, and for Jones to have exceeded the authorized amount by £500 was too much for them to let go unanswered. They felt the company had quite enough to cope with because of arrears due on land sale installments, without advancing further monies for projects which local inhabitants should undertake.

Jones continued to be vulnerable to criticism, sitting as he was in Goderich. Colbornites John Galt Jr., Daniel Lizars, Henry Hyndman, and others including Tuckersmith Township resident Henry Ransford,[54] who supported the Colbornites on occasion, had sent off a list of grievances to the directors on such matters as roads, taxes, the steamboat, sales policy of the company and the harbour. The directors had intended to answer their criticisms fully after securing the necessary background from Jones. Had he complied with their request, Jones would have improved his own position with the directors and would have allowed them to take an objective and detached view of the issues before answering the complaints. Jones, however, could not resist getting into the fray, much to the annoyance of the directors:

Elizabeth Mary (Strachan) Jones with young boy, circa 1845. She married Thomas Mercer Jones in 1832. They had four children: William (died in infancy), John Strachan (1835-1836), Strachan Graham (1838-1869) and Charles Mercer (b. 1841). The young boy is presumed to be Charles Mercer. St. Mary's (near Stratford) is named in honour of Mrs. Jones. *Courtesy of the Trinity College Archives.*

> ... they [the directors] should have considered it particularly inexpedient that it should have led to a discussion in Canada between yourselves and the Settlers.[55]

Jones, of course, was not on favourable terms with the Colborne Clique who by this time had managed to secure most of the better government jobs in the district. He and John Galt Jr., who was now a magistrate and collector of customs, were particularly antagonistic towards one another, and Jones' refusal to pay a customs account owing did little to lessen the bad feelings:

> Enclosed is the account I said I would send you. In it you will perceive that I have stated that you decline to pay the duty on the Chimes and that you deny having received the Pork .... The duties on these articles I have constrained officially forthwith to pay the Government only regretting that from a wish to oblige I should have placed myself in the foolish position of having been duped.[56]

He continued that he had not charged duty on an almanac that Jones had purchased, having preferred to pay it himself rather than having to discuss the

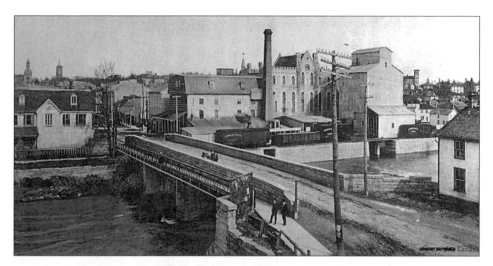

St. Mary's, 1909. The original site of St. Mary's was called Little Falls,
but the fledging community was seeking a new name. In 1845, the wife of Canada
Company commissioner T.M. Jones, Elizabeth Mary Jones, suggested "St. Mary's."
She offered ten pounds for a school building if accepted. The photo
shows Victoria Bridge, flour mill and Opera House, with the stone water tower
and Town Hall tower in the background. The apostrophe is no longer used in the name.
Courtesy of St. Marys Museum.

matter with the Canada Company commissioner. He concluded his letter by
stating: "I do not believe any Gentleman in Canada or elsewhere would have
acted as you have done."[57] This was not a particularly good sort of relationship
to have with a magistrate and collector of customs anywhere, but particularly
so in a small community in a still untamed part of the country.

Despite criticism from headquarters, it seems that Jones continued to be on
reasonable terms with the directors, and so one can only surmise that he was
performing his duties more or less to their satisfaction. It would be another
seven years before he would be released from the company's employ. His marriage to Bishop Strachan's daughter may well have influenced the company's
decision to keep him on.[58] Frederick Widder, in the meantime, was garnering
support from the directors and was even being praised by certain settlers in the
Huron Tract. Whether the support from the Huronites was genuine, or
whether it was merely to "show up" Jones is difficult to discern. In June 1840,
"A Huron" wrote to the *British Colonist*:

> ... we sincerely rejoice to learn, that the intelligent Mr. Widder
> has already brought on a vast change for the better in the
> manner of conducting the affairs of this Company, particularly
> in Toronto ... He ... does much to remove from the public
> mind the unfavourable impressions which the past mismanage-
> ment of their affairs has in many instances given rise to.[59]

Widder was well-connected in Toronto and had won the respect of both the
government and business community – and interestingly enough, the Reformers
liked him too. Being a self-starter, he wrote to the directors on such diverse
subjects as the price of wheat in the colonies and its relationship to the price of
land, new schemes for encouraging emigration and suggested methodologies
for collecting bad debts. Perhaps his greatest single success was the leasing plan
that he developed in 1841.

In suggesting his plan to the directors, Widder had been acutely aware of
the problems facing the provincial government, and more particularly the
Huron Tract (or "the Huron," as he called it).[60] The Huron, to this point, he
said, had been settled almost entirely by employing settlers to make roads and
carry out other improvements. These projects were gradually winding down, or
had not continued in proportion to the increasing population. Few new settlers
were now being attracted to the Huron, and the district was increasingly being
left to its own resources. While he pointed out that considerable numbers of
emigrants had been induced to settle in the Huron Tract because of relatives
and friends, this source was also being exhausted, especially as the settlement
was not now offering the same inducements it had been doing. The Huron was
now dependent on the natural increase of its population. It was remote, he
pointed out, and took time, trouble and expense to reach.[61] Furthermore,
company prices for land in some townships were above the general market
price, but he had reason to believe that many settlers in the older settled parts
of the province were anxious to move westward to the newly opened lands so
that family groups would be able to stay together. In the older regions, wild
land was now scarce and what land there was available could only be purchased
at prices beyond the means of the average settler. That said, prospective settlers
were prevented from moving because of the difficulty of disposing of their
improved farms for cash. To meet these difficulties, Widder developed the
leasing plan.

English-born Frederick Widder (1801-1865) was appointed a Canada Company commissioner in 1839. Extremely hard-working, he was based in the company's Toronto office. He developed a lease-purchase plan which was hugely successful. *Courtesy of the Toronto Public Library (TRL), J. Ross Robertson, Collection: JRR 1002.*

The original plan was intended for the Huron Tract only and involved renting the land at a progressive rate over twelve years, beginning at 5d per acre the first year and ending at 3s/4d the twelfth year. As amended, the settler was to pay over a term of ten years, with no money down, and the plan was extended to all the company lands. The rent would be payable annually, and would be more or less equal to the interest on the value of the land at the time of purchase. For example, if one hundred acres were worth £50, or 10s an acre, and the current interest rate was six per cent, the annual rent would be £3. During any period of the lease, the settler was permitted to purchase the land outright for a fixed advance on an agreed price as follows: if the value of the land was 10s per acre, and paid for within five years, the advance required was 1s/3d per acre. If paid before expiry of the lease, the cost would be 2s/3d per acre. The settler was required to pay all taxes, but would not receive his deed until all monies owing were paid.

In conjunction with his new scheme, Widder suggested the establishment of a settlers' savings bank. He proposed that the company agree to receive any sum on deposit from its lessees for a fixed interest rate of six per cent per annum. The settler would be able to accumulate funds to complete his purchase, and the company would be able to use the money on deposit for its own purposes. Widder saw many advantages to his scheme. Though increased sales, the company would be relieved of some of its heavy tax burden and increased migration to the Tract would create a market for the families already in the Huron – and the existing shortage of labour would be reduced. Furthermore, present difficulties experienced with defaulters or with land abandoned would not occur, because the land would revert to the company if not paid up, and the settler would be encouraged to take an interest in the property because it would ultimately belong to him.[62]

The directors readily agreed that Frederick Widder's plan was "highly desirable" and hoped that with such an arrangement, a new claim "to the character of great benefactors of the province" would be established, by supplying capital to the emigrants.[63] They hoped the arrangement would bring a large influx of settlers to the Tract, along with a corresponding increase in land sales. To be fair to him, Thomas Mercer Jones had suggested a plan of his own, but it had become lost in Widder's much more ambitious scheme. Jones' plan had included a reduced price for land on the back concessions and in those parts of the Huron Tract which were most remote from settlement. He had advocated a price of 8s/9d per acre with a ten per cent discount for prompt payment. While the directors agreed that such a price would allow for increased sales in a shorter time, they could not agree to a discount. They also rejected a suggestion he had made of holding back certain reserve lots for sale in other parts of the province until higher prices could be obtained. To do so would open up the company to criticism, they said, and would expose it to possible hostility. In the end, Jones was allowed to sell land for the lower price if, and only if, he was able to obtain one-fifth of the price as a down payment.[64]

In conjunction with his leasing plan, Widder wrote to the directors about the need to increase advertising. He proposed to direct a campaign toward the older and more settled parts of Canada West, but he also wished to reach European emigrants in the United States during a period of unsettled economic conditions. These had begun in 1837 and were the result of reckless speculation which had led to cotton prices falling by half in the New Orleans market. The ripple effect of this, in turn, led to greatly increased unemployment in New York, multiple riots, inflated prices for foodstuffs and a dramatic fall-off of public land sales from 20 million acres in 1836 to 3.5 million acres in 1838. The effects of the panic persisted in the United States until 1842-1843.[65]

Widder saw much more value in extensive advertising than in hiring agents and commented: "possibly it will not cost more than the Salary of one Agent & his expenses for three months."[66] He did, however, suggest the appointment by the government of a special emigrant agent at New York to induce emigrants arriving there to proceed to Canada. This agent if appointed would, of course, be at no cost to the Company. Widder could also see no reason why the government should not have an agent at both Montreal and Quebec City.

There is no doubt that the company saw the introduction of the leasing plan as an ideal opportunity for increasing the price of their lands. In fact, soon after the plan was introduced, the commissioners were instructed to revise the rents upward so as to exceed the selling price of the land if sold by instalment purchase. Also, after some initial hesitation, the commissioners were advised to allow persons indebted to the company in the Huron Tract to convert land sold by installment purchase to a lease. Frederick Widder agreed with the conversion of old sales to leases, but was opposed to increasing land prices to the extent directed from London. Currently under the old plan, he said, the company was obtaining 11s/10d per acre, while under the lease plan, just three months old, 13s/10d per acre was being charged successfully, including interest, when calculated over the duration of the lease. He emphasized that it would be most unwise to revise the prices upward any further because the government was anxious to promote settlement in the territory immediately north and east of Colborne Township and was selling land at 6s/3d per acre – cash, or to be paid for in labour. Moreover, a new road was being proposed from Owen Sound, on Georgian Bay, to the Township of Ashfield, immediately to the north of Colborne, and free grants of fifty acres were to be given. A road was also planned from Yonge Street to "Saugine" [Saugeen] Bay – at present-day Southampton on Lake Huron. Competition was increasing and a rise in land prices so soon after the plan had been successfully launched could backfire on the company. "The Huron is now in 'Fashion,'" he said, and the company should be extremely cautious not to diminish this feeling.[67]

The directors agreed and delayed considering a price increase. They were critical of the government for opening the area north of the Tract for settlement, however, and viewed it as an "infraction of the understanding with which the Company commenced operations, and one of those circumstances fairly entitling them to consideration by Government on completion of their agreement."[68]

The government would not listen to any company argument in this regard, but Widder had made his point and was listened to by the directors. The probability of increased traffic to the Goderich Harbour was the only consolation for the directors. The hope was expressed that with the expected positive change to the tariff structure of the British Corn Laws,[69] and a reduction in freight charges from Quebec, agricultural production in the Huron would find a ready market in Great Britain. They hoped, as well, that persons engaged in trade would come forward as purchasers at Goderich and other points, and would set up businesses without the assistance of the company.[70]

In sanctioning the leasing plan, the directors pointed out the necessity of punctuality in the payment of rents. This, in turn, meant that purchasers would not be encouraged, or permitted, to take a lease unless they well understood the terms. The company wanted to avoid problems and disappointment. Accordingly, the commissioners were instructed "by every means, publicly and privately," to have this fully understood in all quarters.[71] Jones and Widder lost no time. A typical letter from Jones to a settler who had wished to convert his purchase under the former plan to a lease read in part:

> On sending the accompanying lease the Commissioners of the Canada Company take the opportunity to most seriously request that you will never for one moment forget the terms and conditions upon which you alone hold the Land – and inasmuch as they are the most liberal and easy yet offered, the Commissioners will absolutely insist upon the strictest punctuality in payment of each of the small rents as they become due [under this lease £3/12/9 annually] – and they cannot too strongly impress upon their settlers that the consequences of any default will be immediately followed by forfeiture of the Lease without notice, and the Land forthworth will be open to a new occupant.[72]

Widder, meanwhile, wrote to the editors of many newspapers in Canada West explaining the company leasing policy. All applicants, he emphasized, were required to accompany their applications with either a recommendation from a magistrate or postmaster, or "the name of some gentleman" as reference. This would enable the company to check, if necessary, on the character of the applicant and the probability of his becoming not only an actual, but a desirable, settler. Widder proffered that such persons as canal excavators were unacceptable because of being ill-suited to look after themselves in the bush. Furthermore, he said, the company would not lease more than one hundred acres to any one settler, unless proof was presented that he would be able to use a larger quantity to his own benefit.[73]

Frederick Widder not only won praise from the directors for his leasing plan, but was commended for a new map he published. It contained up-to-date information pertaining to the company and the latest in cartographical information. His objective in issuing it was to have it widely circulated in Europe, the United States and Canada, by selling it at as low a price as possible. In other words, it was to be a form of advertising which prospective settlers would be willing to purchase because of the valuable information it contained. The directors were most impressed with the finished product which was released in 1842. John Perry wrote:

> I am ... to express the thanks of the Directors for this beautiful
> and accurate work containing so much new, and valuable infor-
> mation, which must have cost you great time and trouble.[74]

The directors were more than a little impressed with Widder by the time of the annual meeting in March 1843. All credit for increased land sales was given to him because of the leasing plan he had introduced. They reported to the proprietors:

> It is due to that able and indefatigable officer of the Company,
> MR. WIDDER, to state that this plan was suggested by him, and
> that he has laboured incessantly and successfully in considering
> and completing it in all its details ...[75]

In June of that year, "having taken into consideration the valuable services of Mr. Frederick Widder," the directors increased his salary to £800 sterling per annum, plus an additional £200 in lieu of house rent.[76]

The following year, in August 1844, the *British Colonist* reprinted a Canada Company report from the *London Morning Chronicle* in which governor Charles Franks complimented Jones and Widder on their work:

> The Canada Company are fortunate in having had the assis-
> tance of gentlemen, as their commissioners ... who, by con-
> stant communication with all persons in the province, had

obtained the best information as to the means of promoting the advantage of the settlers.[77]

Franks then specifically singled out Frederick Widder: "In this respect, Mr. Widder especially, had been indefatigable since he had acted for the Company."[78]

Again, extolling all credit to him for the leasing plan, Franks said:

> Mr. Widder, who by his admirable judgment, catious zeal, and constant labour had shown what could be done by a colonization company conducted on sound principles, both for the public advantage, and its own private interests. As the proprietors could not be aware of the voluminous mass of the correspondence, conducted by Mr. Widder, to accomplish what had just been alluded to, it was necessary to dwell upon it, in justice to that gentleman and to explain how the company had acquired the character they now possessed, both in the colony and at home.[79]

The directors had every reason to be pleased with the leasing system. Sales and leases in the year 1843 had jumped to 174,255 acres as compared to sales of 33,655 acres in 1841, and of 65,604 acres in 1842.[80]

Jones Is Embarrassed

Despite periodic references to Jones in company reports and at company meetings, the Goderich-based commissioner was not receiving the attention accorded Widder. It upset Bishop Strachan that his son-in-law was not receiving appropriate recognition from the directors and proprietors for all of his good work. Furthermore, a proposal by Frederick Widder that the company take over the sale of the Clergy Lands for a fee was more than he could bear. A letter addressed to the directors extolling the virtues of Jones while at the same time criticizing Widder and the company in terms of its leasing system, and a further assault under the pseudonym "Aliquis" was the result.

Governor Franks' further laudatory comments about Widder had represented the last straw for Strachan. Franks had stated: "It would be difficult to overrate Mr. Widder's talents and exertions. He is never satisfied with the performance of his arduous duties whilst anything more can be thought of for promoting the interests in which he is engaged."[81] Strachan began his letter by remarking that he felt well qualified to judge on matters in the Huron Tract because of his travels there in discharge of his sacred duties: "I think it a pithy occasion for submitting to your consideration a very few remarks." He said that he had felt inclined to address a letter to the directors for some time because it appeared that they laboured under some strange misapprehension respecting Jones, "otherwise the faint praise bestowed on him in some recent annual reports would not have been tolerated." He continued that Jones knew nothing about the letter nor that he (Strachan) was otherwise acquainted with his standing. That said, he expressed surprise that Jones' services were not more justly appreciated: "Mr. Jones has undergone labours and privations such as no man would unless his heart were in it." It was true, he observed, that Jones had had the benefit of William Allan's counsel and experience, but as Allan never left Toronto, all arrangements and execution fell on Jones. Under Jones, roads were constructed, mills built and all necessary information given to emigrants, but he acknowledged that affairs of the Company had not necessarily been brought "into general notice." He then attacked the leasing system. Calling it far too stringent, he protested that its sole object seemed to be "to grind the poor settlers by covenants." He concluded that although he did not wish to compare the two commissioners, there was no doubt that Jones was certainly the better man. He signed his epistle "John Toronto."[82]

Perhaps the directors felt they were being a little harsh on Jones publicly, or possibly they did not wish to become embroiled in controversy with Strachan. Whatever the case, the report of the meeting of March 26, 1845 merely stated: "The company's business is very ably conducted in Canada by their commissioners."[83] If the directors played down their accolades about Widder publicly, their Toronto-based commissioner nonetheless continued to win praise in company correspondence.

Bishop Strachan, of course, had not been in favour of the company from the beginning and had only been prepared to tolerate it once he knew the Clergy Reserves would remain under church control. He therefore greeted with alarm a proposal by Frederick Widder, sent to the Reverend W.H. Ripley, administrator of the Clergy Reserves, suggesting the company take over the administration of the reserves for a fee. Widder had recognized that no organized system

"John, Lord Bishop of Toronto," circa 1856. John Strachan finally became Bishop of Toronto in 1839 at age 61 (without remuneration at first). He contributed much to Upper Canada, including establishing Trinity College in Toronto in 1851, and as member of the Executive Council. The ultimate Tory, he would have been about 78 when this photo was taken. He died in 1867. *Courtesy of the Toronto Public Library (TRL), J. Ross Robertson Collection: E 4-1g.*

existed for the settlement of the Clergy Reserves and thought his proposal to be the perfect answer:

> It is now proposed to give the Clergy Lands that full benefit of the position which the Canada Company have acquired at considerable pains and cost, and which necessarily brings to them nearly all the applicants for Lands – for intending Emigrants in Europe are as well acquainted as the old Settlers in Canada upon the facilities which the Company affords them.[84]

A number of newspapers supported Widder. The *British Colonist* commented:

> nothing can be worse than the present management of the residue of the Clergy Reserves .... We hesitate not to express a firm conviction that the Canada Company could manage every acre of land belonging to the Crown or Clergy in Western Canada, at a half, if not a third of the cost incurred under the

present system …. It is well known that private individuals can always do their business at a much cheaper rate than the public ever get similar works performed.[85]

Widder's overture on top of the directors' treatment of his son-in-law was too much for Strachan. Observing officially that "the [Church] Society is not in a position to entertain the question,"[86] he launched into a vindictive attack on Frederick Widder and the company under the pseudonym of "Aliquis." Entitled *Observations on the History and recent Proceedings of the Canada Company, addressed in Four Letters to Frederick Widder, Esq., one of the Commissioners*, these published letters dealt with the period from the formation of the company through to early 1845.[87] Dealing firstly with the terms under which the company acquired its land (which he condemned), Strachan went on to discuss the management of the company in Canada under three headings, corresponding to the consecutive administrations of Galt; Jones and Allan; and Jones and Widder. In his first letter, he gave Galt all due credit for his work in Canada. In his second letter, he cited the work of Jones and Allan in glowing terms. Calling them men of sound judgment, he had nary a critical word and praised the company for its colonizing work. It was in his third letter that he began his attack on Widder. "It is not easy to divine why the Directors of the Canada Company should have in any manner disturbed the management," he began. "On your [Widder's] arrival in 1839 you found everything in the most excellent condition, and little employment for a third Commissioner."[88] Strachan continued that when a division of labour became necessary, Widder and Allan were left to handle the simpler and easier portion of the company's business. The more arduous task of extending settlement in the Huron Tract was left to Thomas Mercer Jones:

> Not only, I presume, because he wished to complete what had been so long under his care, but because it was a labour that no stranger could satisfactorily conduct.[89]

For some time, he said, matters went on as usual, and then he asserted in libellous terms:

> To a person of your disposition, so anxious for notoriety and so prone to scheming and maneuvering, this healthy state of

things was by no means agreeable, – no faults to find, no amendments possible .... Something ... must be done; some stroke of policy to render you conspicuous and ingratiate you with the Directors. But how? Mr. Allan was an enemy to unnecessary change; he had little confidence in the judgment of a man who had no experience in the country, and who might be a very good clerk in the counting-house, and yet by no means qualified to conduct the complicated affairs of the Company, or improve a management which had demonstrated its excellence by many years' experience. The first step ... was to get rid of Mr. Allan ....[90]

And so Strachan continued. Widder, he said, had been responsible for stopping works so necessary to the settlement; he had introduced the leasing system, a "system of extortion and rapacity," and like "a covetous trader ... determined to sell every indulgence or extension of credit at the highest rate."[91] Strachan's tone remained unchanged throughout the rest of his diatribe, and he concluded by remarking that the character of the company had never stood high in the province:

> for its policy has ever been selfish, and since you had a share in the management it has been rapidly sinking. The former Commissioners, whatever their inclinations may have been, had no power to do anything that might promote the general benefit of the Province; but as faithful servants and devoted to their duty, they gained a character for plain dealing and honesty of purpose; they acted fair and equitably with the settlers; they admitted of no trickery in their transactions; all was open, fair and above-ground; ...[92]

If Strachan thought he would have the support of the press and settlers, he must have been very disappointed. He was out step with the times and had misjudged the climate of opinion in Canada West towards the company since the introduction of the leasing system. Furthermore, most residents of the province knew little or nothing of the Huron Tract, including the issues there more generally, and those in Goderich more particularly, let alone knowing anything about

Thomas Mercer Jones. The net result was that Frederick Widder rose in stature and Jones was so thoroughly embarrassed that he wrote the directors and denied in the strongest terms possible that he had any knowledge of the publication, or anything in it, until it was shown to him by Frederick Widder.[93] The directors replied that they were perfectly satisfied that he could not have had anything to do with the publication or its preparation. They had been quite convinced of this, they said, without his disclaimer, although they were glad to receive it, "as shewing that he enters fully with them into the injustice exhibited by the writer towards Mr. Widder and the management of the affairs of the Company."[94]

To Widder, the directors expressed their exceeding regret that such an attack could have been perpetrated against him. Company secretary Perry wrote:

> They [the directors] are too well convinced of his merits, which they frequently had the satisfaction of recording, to say more than they consider this work proceeds from some person having interested and unworthy motives in view that it is full of assertions contrary to the fact, and proceeding as it does from an anonymous writer, it is beneath their Serious attention or that of Mr. Widder.[95]

The newspapers of the Canadas featured front page editorial comment on the attack by "Aliquis," the opinions expressed being akin to those of Secretary Perry to Frederick Widder. The *British Colonist* in a byline from the *Montreal Gazette* said it was obvious that the object had been to abuse the company in general, and Widder in particular. The writer of the article said it was not his wish to enter into any of the questions at issue between the company and its lessees and other inhabitants of the Huron District. He pointed out that such disputes were always incidental to the possession of large property, particularly when sold either on a system of credit or reserved rents. The terms of the company may be advantageous or disadvantageous to the settler, he commented, but "they are open and intelligible. No man need buy of them unless he likes …." He concluded:

> … what we wish to remark on is the exceedingly bad spirit in which the pamphlet is written. It is flowing over with malice and all uncharitableness, and all because Mr. Widder as Land

Agent, made a proposition for the rescuing of the public lands of every description from the wasteful management of the land office .... To revenge this heinous offence, everything is done to damage both the Company here and at home. We can conceive nothing more mischievous in any country than the prevalence of such a spirit directed, as it is in this case, to accumulate odium on individuals, to shake the security of the property, and to deter capitalists at home from making investments in the colony.[96]

Company Victory

Public opinion came to the defense of the company over the "Aliquis" attack and would do so once more, but by a much narrower margin and over a much more provocative issue. Tiger Dunlop's bill in the legislature in 1845 to legalize an illegal bylaw of the Huron District Council, in an attempt to recover company back taxes, was the end result of a feud which had gone on between the company and the Huron District Council for some six years.

The quarrel between the company and council over the tax question dated back to the loan of £3,100 made by the company to the District Council to build the gaol and courthouse at Goderich, and to the Municipal District Council Bill of January 1, 1842. From the beginning, the Huron District had been short of money, and so the company had agreed to advance the needed funds on condition that the District Council allow the company to hold back an equal amount of its taxes, including the yearly instalment and interest. The council accepted this arrangement reluctantly, but almost immediately, the magistrates tried to force the company to pay all of its taxes with no holdback. This the company refused to do until the issue was settled, and so was incensed when the council attempted to extort money by applying a tax on the company at an excessive rate under an illegal bylaw.[97]

The Municipal District Council Bill had empowered all district councils to levy property taxes on the basis of a previously drawn-up financial statement, indicating expected receipts and expenditures. Despite the bill, the council, made up of such Canada Company foes as Daniel Lizars, John Galt Jr., Henry Ransford and Henry Hyndman, proposed to levy a tax under a bylaw without striking a rate, and to tax unsold or wild land of the company at the same rate as land under cultivation, namely, one penny an acre.[98] Clearly, the company

Charles Widder's residence, Goderich. Charles Widder was Canada Company
commissioner Frederick Widder's much younger (by 18 years) brother.
He worked for the company until the Goderich office closed in 1852 and
then became a Crown Land agent. Charles built this home overlooking the
harbour on Waterloo Street in the 1840s. *Courtesy of Cayley Hill.*

was being targeted because it owned most of the wild land in the district.

The directors thus had good reason for advising the commissioners not to
pay the tax. Instead, they instructed the commissioners in September 1844 to
effect payment under the old rate in an attempt to break the deadlock. The
amount of £3,813/10/5 was then tendered without prejudice to any claim by
the district and without reference to the £3,100 loan.[99] This was the second
time in the last four months that the directors had attempted to pay the tax:

> In adopting this course the Directors are not influenced by any
> love of popularity in the Province, which however desirable it
> is in itself, can only be valuable if obtained by measures of strict
> justice, and regard for the rights of others.[100]

The District Council had been steadfastly refusing to accept the company's
payment and "Tiger" Dunlop went so far as to order the tax collectors in the
townships to accept no taxes for 1843. If the company would not pay under

the new rate, nobody should, he decreed. The next year, many of the taxes collected were refunded for the same reason. By 1844, the council had run out of
money. Roads were deteriorating and district employees were going unpaid. In
an attempt to portray the company in the worst possible light, the council even
allowed their office furniture to be repossessed for non-payment.[101]

In January 1845, Dunlop introduced a bill into the legislature entitled "For
the Recovery of Certain Taxes in the District of Huron." In effect, he was
attempting to legalize the assessment bylaw and, furthermore, he proposed that
the company pay a penalty for disobedience once the bylaw was passed.

While Dunlop was lashing out at the company in the legislature, Frederick
Widder was busy preparing a memorandum for members of the legislature to
which he attached Dunlop's *Defence of the Canada Company*, written some nine
years earlier while still in the employ of the company. Unfortunately for
Widder, his paper did not reach Montreal until after the third reading and the
bill was carried fifty to twelve. The solicitor general for Canada West, and the
provincial secretary had both voted against the company. Widder was dumbfounded and wrote to an influential member of the legislature from Canada
West, William Hamilton Merritt:[102]

> I was well aware of the *prejudice* that existed, but would scarcely
> have imagined that they should have been made public *officially*.[103]

The bill had still to be put before the Legislative Council and Widder lost no
time in working behind the scenes to gain support for the company in the
meantime. He sent a copy of his *Memorandum* and Dunlop's *Defence* to all the
newspapers and said to Merritt:

> this Document will no doubt both surprise you and make you
> laugh – but yet can I think be made most effective by you with
> your friends.[104]

Widder said he could not imagine that the bill would be passed by the
Legislative Council who "are supposed not to be led by popular prejudices,
or *any kind*, of *influences*, but those of justice & honesty."[105]

Widder's efforts and Merritt's influence had the desired effect. He was able
to report to the directors that the bill had been thrown out by the Legislative

Council in a close seven to six vote. The company's last-minute success had been due to one factor, namely, its support from the colony's financial community. The *Montreal Herald* reported:

> There is not a question in our minds, that if that bill had passed into law, a blow would have been struck at our [the colony's] credit in England, which would have been felt in all our transactions. It ought to be our duty, as it is our interest, to follow in the footsteps of our parent land, and maintain our credit, by preserving with jealous care the rights of property conferred upon individuals and companies. It is only by such conduct that we can attract to Canada the plethora of capital which exists in Britain, and which is every hour seeking new fields for investment.[106]

The directors were jubilant at the company's last-minute success, and Frederick Widder saw some good as having emerged from the fight:

> if any doubt did exist upon the motives which directed & ruled their previous attempts against the Company these doubts will now be removed – the instigations of the difficulties of the Company be no longer unknown ....[107]

He added that the District Council, upon the presupposed passage of the bill in the Upper House, had determined to avail itself of the precedent established. They were going to require the legislature to grant it power to tax the wild lands at 2d per acre over and above the 1-1/2d per acre presently allowed under the Municipal Act. While Widder told the directors that the Council had been humiliated, he said he feared that the company lands would be taxed as heavily as possible. He recommended therefore that the company should sell its lands as quickly as possible, even to the extent of allowing free occupancy in the rear concessions for a period of five years subject to certain improvements being effected yearly. At the end of the five-year term, the settler would then have the option to purchase. If he should decline to do so, the settler would be allowed the value of his improvements by the next settler who would purchase it. Widder had recommended this plan, not only because of the tax load which would have to be borne by the company, but also because there had been a petition before the legislature to lower the price of government land in Wawanosh and Ashfield townships north of Goderich. A long period of credit had been proposed and furthermore, all the Clergy Reserves were to be opened for sale immediately at one-third down with the balance due in four annual instalments.[108] The directors chose not to adopt his proposal, and while sales and leases did dip considerably in 1846 to 48,627 acres, they rose again in 1847 to 117,940 acres.[109]

On defeat of the bill, Canada Company governor Franks personally wrote a joint letter to the commissioners, and expressed his thanks for their exertions "in repelling the injustice that would have been inflicted upon the Company." Although Thomas Mercer Jones and Frederick Widder both received recognition from the governor, Jones would not have been happy to read in the same letter:

> Resolved: That it be recommended to the Court to request Mr. Frederick Widder's acceptance of the sum of £200, as some return for his great exertions during the late session of the Legislature of Canada, in resisting the Huron District Taxation Bill.[110]

Harmony Between Company and District Council

With the defeat of Dunlop's bill, the company and the Council went to considerable lengths to reach an agreement over the tax issue. At least they were now able to co-exist on reasonable terms. Jones was even considered by the Council for the position of warden in the District – but would he get drawn into politics once again? The company began the collection of arrears again in earnest and Widder was able to negotiate a final settlement with the government for the lands purchased under the terms of the charter.

The directors would make the first move to restore harmony. They had, of course, been perfectly willing to settle the tax issue under the old bylaw, but suggested that perhaps some compromise would be appropriate. They even expressed a readiness to not insist on the immediate payment of installments due for the Huron gaol and courthouse.[111] It would not be until June 1846, however, before a compromise was reached. Under the agreement, the company paid £6,455/8/7 in back taxes from July 1, 1842 to December 31, 1845 (the bulk of the disputed taxes prior to that year had actually been paid in June 1843). The Council in turn paid the company £2,660 covering seven loan instalments, including principal and interest.[112] A reconciliation between company and council was finally taking place and, in 1846, Jones was indeed offered the wardenship of the District of Huron.[113] He accepted, and the directors expressed the hope that his appointment was proof of the general desire that the affairs of the district should proceed harmoniously. They also hoped that as warden, he would find it easier to obtain repayment of the district loan and the advances made to the district officers.[114] Two-and-a-half months later Jones resigned his office.[115] Perhaps he found having to work so closely with his old foes just too uncomfortable.

Jones may have resigned this office for another reason. As warden, he was a servant of the district residents. Yet, as company commissioner, it was his duty to press for payment arrears from many of those same people. Clearly, the two jobs were not compatible, particularly when considering the recent past history of the area. If he wished to compete with Widder for laurels from the directors, he had to press for payments – successfully – and to increase sales. Pressure was being applied from London:

> The Directors can see no ground whatever for submitting to the delay in payment which has become so general with those indebted to the Company in every part of the Province; and in the Huron District in particular.[116]

A typical letter to a settler in default read:

> Canada Company's Office
> Goderich 1st Decr. 1846
> Sir
> The various Letters addressed to you year after year for such a length of time by the Company's Commissioners expostulating and threatening but without actually adopting harsher measures must have satisfied you of the extreme reluctance which they have felt to harrass or put you to the heavy expense of Legal proceedings in pressing for the payment of the arrears due them for the many years past. – Their patience is now however exhausted and looking at the extent of the Improvements you have made on the Land you hold under the Company and the

value of your Stock whilst at the same time you have never paid one Shilling on account of the Land it is very clear that a continuation of that indulgence is no longer an act of Kindness, since you have ample means of paying off a considerable portion of the arrears you owe the Company if not the whole.

I have therefore to inform you that we have most reluctantly come to the determination of placing your account in the hands of our Solicitors for collection and you will thus incur a heavy expense which a little exertion on your part would have avoided.

I am, Sir

Your Obt. Sert.

Thos. Mercer Jones

Comm.[117]

Whether the settler paid his arrears of £128 (which included £89.4.4 in principal and the balance in interest) voluntarily or whether the account was put in the hands of the company solicitor is not known. What is known, however, is that a serious attempt was finally being made by Jones to collect arrears in his territory. He reported to the directors that everything short of coercion on his part was being used.[118] Why, though, had he not been keeping a closer tabs on overdue accounts earlier on? The settler to whom he sent the above-noted letter had purchased his land some nine years earlier and yet had been sent only one notice of payment in arrears.

By June 1846 it is obvious that the directors thought of Frederick Widder as the senior commissioner. Evidence of this is seen in a letter addressed to Jones and Widder over the advisability of relaxing the director's orders to reserve a strip of land on either side of the proposed railway from Toronto to Lake Huron. Widder had been in England the previous winter in an endeavour to raise funds for the proposed railway from Toronto[119] to Stratford through Guelph, and thence either to Goderich or Sarnia (the directors favoured Goderich as the terminus with a branch line from Sarnia to Stratford).[120] So enthusiastic had the company become over the railway that they instructed Jones to reserve land on either side of the roads from Guelph to Stratford, Stratford to London, and from Stratford to Goderich (of course, this was not possible, given that the company did not own a continuous strip of land beside these roads). He was also to hold from sale any unsold town and park lots in

Goderich, Guelph, Stratford and Mitchell. Likewise, all lands owned by the company containing water power sites within twelve miles of the above roads were to be held off the market.[121] While Jones had been instructed to do this as quietly as possible, it became obvious that hostile feelings were being aroused by the policy. He requested a relaxation of it. The directors replied:

> The Directors leave it to him [Widder] to decide what course you should adopt in the disposal of lands ordered to be reserved.[122]

With the District Council and the company working together more amicably, Jones was able to concentrate on the business of the company at hand, i.e., selling land in the Huron Tract and collecting payments. He was instructed, moreover, to have no part in the proceedings of the District Council at Goderich, nor in the discussion surrounding the separation of the five eastern townships (Ellice, Downie, North Easthope, South Easthope and Wilmot) from the Huron District.[123]

> The Directors are desirous of taking no part in such divisions as far as they can prudently abstain. They are anxious to throw upon the inhabitants the duty and responsibility of self-government, and would be glad to see more energy on the part of their settlers generally. Hitherto the practice seems to have been to throw everything upon the Company, instead of considering them interested as part proprietors only, in the general welfare of the District.[124]

All taxes, he was told, were to be paid promptly, and a reciprocal spirit was expected from the council in payment of the loan.[125]

In September 1847, Frederick Widder arranged for the final release of land from the government under the charter and subsequent amendments. In return, the company gave a bond guaranteeing to spend the approximate £3,000 remaining in the Huron Tract Improvement Fund.[126] At the conclusion of the negotiations, Widder was once again singled out. Secretary Perry wrote:

> I am particularly to express the strong sense they [the directors] entertain of the important service rendered to the Company by

> Mr. Widder in his able and persevering exertions in bringing
> this business to a conclusion.[127]

The directors were extremely pleased that the company was finally free of its commitments. Although overdue accounts amounted to £170,000, and arrears in rents had reached £35,000 by November 1847, the company was now in a sound financial position. A surplus of £123,287 was shown at the end of 1849 and of a total of 907,888 acres of land were left for disposal throughout Canada West. Since 1838, all management expenses and shareholder dividends had been met from collections in Canada.[128]

Thomas Mercer Jones' Dismissal

By 1850 Jones ran company business in the Huron Tract with increasing independence – and when he was in contact with the directors and Frederick Widder, he was on the defensive. He clearly resented Widder, and fell in more and more with the "locals." As noted earlier, his actions precipitated his downfall.

The dismissal that year of an office employee, a "Mr. Robertson"[129] for "defalcations" was the first sign of that increasing independence. Jones casually told Widder about the firing and let the matter drop. When the directors heard unofficially about it through Widder, they were indignant[130] and after expressing their frustration to Jones, informed him that the appointment of a replacement was to be a temporary arrangement only. Furthermore, from here on in, they said that as much business as possible was to be transacted in Toronto because consideration was being given to downgrading the Goderich office to agency status. Like other company agencies, the office at Goderich would merely effect sale preliminaries, with final processing to be completed at Toronto.[131]

There seems little doubt that the company would have closed the Goderich office sooner had Jones not been the incumbent. He had been in the service of the company since 1829 and could not really be released without just cause. However, given the animosity between Widder and Jones, a move back to Toronto would have been out of the question. Furthermore, his marriage to Elizabeth Mary Strachan would have been a consideration as the directors decided what to do next.

Jones must have felt uneasy these days because on his return to work after an illness in the fall of 1850, he forwarded a medical certificate to London to prove he had been laid up. Governor Franks personally wrote to Jones and said his own statement would have been perfectly acceptable to satisfy the directors that "you would not have discontinued the exertions with which you have laboured so long, and so energetically for the Company, without a strong necessity ...."[132] If Franks' letter satisfied Jones, a further increase in Widder's salary of £300 (for a total of £1,100) which was advised in a joint letter some four months later must have really hurt. Since Widder's appointment, the Toronto-based commissioner had received three raises, plus a £200 one-time bonus and a £200 annual housing allowance, yet Jones had apparently received neither raise nor bonus in fourteen years. The latest notification from Perry of Widder's increase read:

> I am desired by the Directors ... to express to Mr. Widder their
> strong sense of admirable manner in which he has conducted
> all the affairs of the Company, under his charge.[133]

It is apparent that the directors had still not decided if and/or when Jones was to be eased out when they wrote to him in October 1851 about arrangements for consolidating the Goderich and Stratford bank agency balances at Toronto under Widder's control.[134] It is possible to see a pattern emerging, however. The directors continued to discover errors in Jones' work and they began to question his actions much in the manner used some twenty-four years earlier when Galt was being eased out. They questioned Jones' tardiness in not depositing receipts from sales and leases in the bank immediately for the period September 1850 to July 1851, and demanded a full explanation. They queried his not spending the small sum of £14/7/9 at the harbour in the fall of 1850 in order, as Jones had said, to prevent serious damage in the spring breakup the following year. Jones tried to explain his failure to undertake the repairs by citing, besides his illness, a directive from London instructing him to spend no more on the harbour. Perry replied:

> They [the directors] are aware what the state of Mr. Jones's
> health was, they sincerely regret to say, a sufficient reason for
> the want of attention to this matter, otherwise the Directors
> cannot consider that he was justified, for such a trifling expense

... in leaving the whole work at any serious risk of destruction; nor in considering the expressions used in the Court's letter of the 18th July last, (which were only a renewal of the opinions so often given before, that they would not make any general or considerable repair, or extension of the whole work,) as precluding you from such an expenditure as that referred to, whereby, perhaps the whole work might have been saved from destruction, or serious damage.[135]

The harbour expenditure incident reflected a mounting frustration with Jones. The directors were now obviously building a case around it for their own purposes.

Not only had Jones raised the ire of the directors, but he had also antagonized Frederick Widder. Having felt uneasy about his latest reprimand, Jones tried to implicate Widder by blaming him for not replying promptly to his letter in connection with what Jones regarded as necessary repairs. While the harbour had been inspected in July 1851, Jones had only notified Widder in late September about the repairs needed and the suggested remedial work. He then demanded an immediate reply. As Widder had not responded by return post, Jones considered that he now had a perfect scapegoat. Widder answered Jones unequivocally:

I am in receipt of your letter ... upon the repairs of the Harbour Works. You urge no fresh arguments in support of the position you are desirous to establish, you merely reiterate your endeavour to throw upon me a responsibility created solely by yourself ... had you not retained from me the result of your examination of the Works of the 30th July, for 57 days, or had upon choosing to do so, taking upon yourself the expenditure of the very trifling sum of £14-7-6, to save from destruction the costly works, you would not now have had to discuss ... whether it was imperative for me on the 26th Sept by *return* of Post to have answered your letter of the 9th August that you might make the repairs in Sept. what intelligent artizan you had engaged had promised & should have done in that month and which you stated there was no denying the fact.[136]

By April 1852 Frederick Widder and Thomas Mercer Jones were scarcely tolerating each other. Jones had wanted to launch an open attack on W.H. Smith's *Canada: Past, Present and Future* because of criticisms leveled at the company. Widder advised Jones that such a course of action would only serve to benefit Smith:

> … there are an abundance of people who having leisure time, and inclination would rejoice in the opportunity to get the Company into a paper war. The mode by which I desire to refute Smith's attack is *not* by alluding to *him* or the *Canada Company*, or its *Agents*. Before you receive this letter I shall I believe, have done all I propose doing, and I hope without feeling much the absence of the information I might possibly have found of some little assistance and which you declined to furnish upon a misconception of my motives.[137]

He continued: "I will not participate in your feeling for immediately I have accomplished my object you shall have a copy of what I had done." What Widder did has not been ascertained, but the correspondence indicates the degree of the strained relationship between the two men. Then there was Jones' refusal to give Widder a copy of the proposed reply to Smith (Widder had passed on his copy to the directors without noting certain figures used by Jones). Jones refused to give him the requested copy, which in turn explains Widder's sarcastic comment. The foregoing was merely the preliminary to an event which would precipitate Jones' release. If the directors were looking for a reason to dismiss him, they would not have long to wait. Jones' blunder of July 1852 not only moved the directors to fire him, but caused the company great embarrassment as well.

The proposed railway to Lake Huron was the central issue. Since the railway had come up for serious discussion, the directors had consistently favoured a line from Toronto to Guelph, and then on to Goderich.[138] In order to ensure that the line from Toronto to Guelph would be built, the directors had even expressed their willingness to assist as agents in the sale of the debentures for the Toronto and Guelph Railway Company.[139] They also had authorized the commissioners to advance £500 towards the cost of surveying a line from Guelph to Goderich.[140] However, while the directors and Widder were actively engaged in promoting

Seaforth Main Street. Laid out by the Canada Company, Seaforth really got its
start with the opening of the Buffalo and Lake Huron Railway in 1858. A thriving
town of 3,000 by the 1870s, Seaforth was incorporated in 1874. In 1876, there was
a devastating fire that destroyed much of the downtown. Many of the buildings
in this circa 1900 photo date from after that fire. *Courtesy of Ken Cardno, Seaforth.*

this railway line, Huron residents were pressing for gravelled and plank roads, a
position which Jones' supported. They expressed little interest in a railway from
Guelph to Goderich while local roads needed attention, and would not contribute towards the cost of debentures, nor contribute to the cost of a survey. In
a letter to Jones, Widder expressed considerable annoyance at this turn of events:

> I have perused your remarks upon my enquiry to whether the
> County Council or Township Municipalities would contribute
> in Debentures or otherwise towards the cost of [the] survey of
> [the] Railway from Guelph to Goderich, by which I perceive
> that you are all of opinion that the Gravelled Roads are of
> primary importance, the Railway secondary, and that therefore
> until the former is secured you will not assist towards the con-
> struction of the latter. I regret much this indifference towards

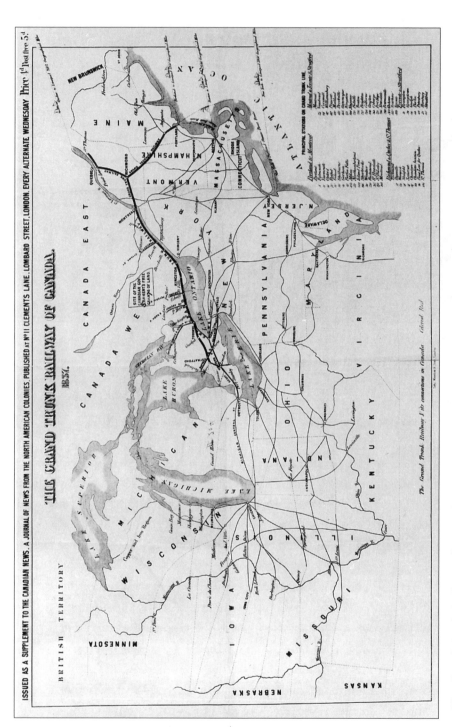

The Grand Trunk Railway of 1857. There were expansive plans afloat for railways in Canada. Optimism sometimes got in the way of reality as railways began a period of unfettered expansion. *Courtesy of the Archives of Ontario, F 129 Pkg 18 #204 AO 6483.*

the Railway, because there is a great risk that it will result in
taking the line in a different course to that which I have been
endeavoring to secure, there are difficulties enough to overcome in securing a terminus at Goderich, without the apathy,
and in my mind mistaken views upon the vastly superior benefits to be obtained by ... Railways over gravelled Roads....[141]

What Widder had not realized was that a group in the Huron Tract area
who favoured railways was actively promoting a line from Buffalo to Goderich
via Brantford as a shortcut for American trade to/from the west. The Goderich
promoters, for their part, saw a railway from their town to Buffalo as a way of
increasing freight traffic through the Goderich Harbour, along with ancillary
spin-off business. To the Buffalo promoters, increased freight traffic through
Goderich was complimentary to what they wished to achieve as they sought to
expand their trade area and encourage use of the Erie Canal which ran from
Buffalo to New York City. Brantford, for its part, had been by-passed by the
Great Western Railway and was looking to be linked to another railway. To this
end, the Buffalo & Brantford Joint Stock Railroad Company had been incorporated in 1851, but the name was quickly changed to the Buffalo, Brantford &
Goderich Railway Company (BB&GRC) in 1852 when the Goderich interest
gelled. To add to the mix, John Galt Jr. of Colborne Township became president of the BB&GRC.[142]

When the directors learned of Jones' active involvement with the promoters
of the BBGRC, they sent a letter to both commissioners which should have
removed any doubt from Jones' mind as to which line the company favoured
and which line the commissioners should be supporting:

> The Directors are of the opinion that the Guelph & Toronto
> Railway would be more for the advantage of the Province and
> the Canada Company in particular, than the line by Brantford
> & request that any steps you take should be for the promotion
> of the former.[143]

Whether Jones had misunderstood a statement by Widder by noting that
there was reason to believe that the company would assist the District Council in
raising £80,000[144] to gravel the roads and that, by extension, the Council

Buffalo and Lake Huron Railway in the early days of the railway. This railway project was controversial from the beginning. Thomas Mercer Jones' support of this line over one from Toronto was the "last straw" for the Canada Company directors. In any case, the directors worried that if the Grand Trunk Railway was completed to Sarnia (which would compete with the Great Western Railway) Goderich would suffer. It did. *Courtesy of the Huron County Museum.*

would support a railway to Goderich, is difficult to say. Whatever the case, Jones presided over a meeting in Goderich of the committee promoting the Buffalo-Goderich line. At that gathering he stated that he was fully justified in believing that the company would render the same assistance and facilities for raising capital by way of negotiating debentures for the Buffalo-Goderich railway line, as the Canada Company had "so readily and generously done for the Toronto and Guelph Railway."[145]

The directors were dumbfounded. How could he have possibly made such a statement:

> ... he must be well aware from the correspondence with the Court for many months past that they had thought it right to do all in their power to promote a line from Toronto to Guelph, believing that such line would necessarily be extended to Goderich, and being strongly impressed with the opinion that an attempt at competing lines would end in a failure in obtaining any.[146]

Jones was told by the directors that he had not only acted inadvisably in presiding over a meeting to promote a rival project, but even more so because of expressing an opinion for which he had no authority whatever, i.e., the negotiation of debentures for a rival line.

William Benjamin Robinson, born in 1797, was the youngest brother of John Beverley Robinson and Peter Robinson. He was appointed Canada Company commissioner in 1852, following the termination of Thomas Mercer Jones. An elected politician, he held many positions in government and negotiated major treaties with Native Peoples. *Courtesy of the Toronto Public Library (TRL), J. Ross Robertson Collection, T30790.*

The directors had been placed in a very awkward situation. If they did nothing, it would appear they had sanctioned his statement. Furthermore, it would also appear as a gross breach of faith on their part after pledging themselves to the Toronto and Guelph Railway Company. Accordingly, Jones was instructed to write a letter to Frederick Widder stating that he (Jones) had no authority whatsoever to make the announcement about potential support for the Buffalo to Goderich line. He was to take care in writing the letter so that it could be used in any way Widder saw fit "to destroy the false and injurious inference which could be drawn."[147] Jones tried to explain himself, but it was too late and in a letter dated November 11, 1852, the directors announced that he was being "superseded." Widder was to become chief commissioner at a salary of £1,300 per annum and was to be assisted by a second commissioner, William Benjamin Robinson, at a salary of £600 per annum.[148] Jones was granted a pension of £400 per annum for life.[149]

With the appointment of W.B. Robinson,[150] the company had a commissioner who was well-qualified in his own right. The youngest brother of Peter and John Beverley Robinson, he was born in 1797 in Kingston and was brought up and educated by his mother in Newmarket, following the death of his father shortly after his birth. In 1822, he married Elizabeth Ann Jarvis, the daughter of the provincial secretary, William Jarvis. He went into business while in his twenties, including fur trading in the Muskokas with his brother Peter. In 1828, he contested the first election in the Legislative Assembly of Upper Canada for Simcoe County, but lost by nine votes. He then ran again in 1830 and won, and was re-elected in 1834 and 1836. He lost the election to the first Assembly of the united Canadas in 1841, but won in 1844, and was re-elected, or won by acclamation, until the election of 1857 (for the 1854 election he ran in Simcoe South, the constituency having been split the previous year). Concurrently, he was appointed co-commissioner by the government of Canada West to superintend expenditures for improvement of the Welland Canal (1833 to 1843). He also negotiated treaties with the Chippewa First Nation of the Lake Simcoe area in 1843 for land which was to be set aside their use. In 1844, he was appointed inspector-general for the province (with a seat on the Executive Council, but he resigned from the Executive Council on a matter of principle the next year). Two years later he became the chief commissioner of public works and negotiated two claims in 1850 with the Native Peoples. One, with the Lake Huron Chippewa band (the Robinson–Huron Treaty), involved a huge area running north from the north shore of Lake Huron "to the height of land" and over to the Quebec border, while the other, with the Chippewa of the Sault Ste. Marie region (the Robinson–Superior Treaty), comprised a very sizeable land area running east, north and west around Lake Superior "to the height of land" from the border of the Robinson–Huron Treaty on the east to the Minnesota border on the west. As a member of the elected Assembly, Robinson opposed the secularization of Clergy Lands and took an active part in proposals to construct an intercontinental railway linking Canada and the Maritimes. A consistent supporter of the British connection, he shared the "Robinson charm" – a sense of humour and a zest for living. With his evident energy and ability, and being well-connected to the seat of government, he would be an excellent counterpart to Frederick Widder. The company chose well.

The announcement of Jones' dismissal and the later decree that the Goderich office was to become an agency only caused much ill-feeling in the town. Former adversaries of Jones sprang to his defense and libellous statements were thrown at Frederick Widder and the company. So vindictive were the statements, in fact, that the directors felt obliged to come to Widder's defense and state precisely to him the reason for Jones' dismissal:

> ... they [the Directors] feel bound, when observations are made
> on your conduct, which might leave painful impression on your
> mind, or those of your friends, to state for your satisfaction that
> has led to the recent changes. From the experience the Directors

have had of the Company's affairs since you have acted as a Commissioner, they are perfectly convinced that the present prosperous state of these affairs must chiefly be ascribed to your able and judicious conduct, and they consider it of the greatest importance that you should be enabled to apply your unfettered exertions thereto. They have observed that on certain occasions for years past, some of your most important measures seem to have been interfered with in a manner calculated to add to your difficulties in performing your duties; but never so obviously as in a case which recently occurred, and which convinced the Directors they were no longer justified in allowing the affairs of the Company to remain subject to the conduct of the Two Commissioners having equal authority, but that it was requisite that you should be appointed First Commissioner ....

... it is a duty to you to state the facts as they *are*, and to enable you if you think fit to explain upon sufficient authority why the recent changes have been made.[151]

The directors were obviously dismayed with Jones and had decided it was time for him to go.[152]

After the dismissal of Thomas Mercer Jones, company correspondence between the directors and commissioners was reminiscent of 1829:

In regard to the numerous errors which appear to exist in the business conducted at the Goderich office ....[153]

Furthermore, it was discovered that Jones had appropriated for his own use the property immediately north and west of the Goderich office to the waterline (most of the present Harbour Park down to the site of the present-day grain elevators). This had been done in spite of company policy which forbade an employee to deal in, or to purchase, company land without approval of the directors. Having become aware of Jones' action, the directors demanded the surrender of his claim, but it was not until 1855 that he complied and returned the property to the company.[154]

A very legitimate question comes to mind: Why did the directors keep Jones in the employ of the company for as long as they did? Two reasons present

Lighthouse, Goderich, circa 1860. Situated 120 feet above Lake Huron overlooking
the harbour and Maitland River mouth, this replaced an older outer "range"
tower which had been built further back on Cobourg Street (on the site of the
Bank of Upper Canada). Built circa 1851, the lighthouse has been in continual use
since then. The ancillary buildings were removed some time after 1913.
Courtesy of Duncan and Linda Jewell.

themselves. Firstly, Jones was conscientious, if misguided at times, and had been
of invaluable help to Frederick Widder when Widder was first hired. Indeed,
Jones was regarded as senior commissioner at that time. The 1830s had been a
difficult period and Jones had performed reasonably well, all things considered.
With William Allan as his counterpart resident in York/Toronto, there was a
seasoned professional to act as his "sounding board" on the many issues of the
day confronting the company. But as Allan was busy with various other pursuits
as well, it was Jones who kept things going on a day-to-day basis during that
period. Furthermore, as has been noted, Jones' connections and status through
his marriage to Bishop Strachan's daughter, Elizabeth Mary, could be helpful to
the company, so from that standpoint, it was certainly an advantage to have
Jones in the company employ. With Widder's arrival, though, Jones now had a
direct "competitor" who would be devoting full time to company business.
Widder was an amiable and practical type of person with "business smarts" and
administrative talents. Expansive in his thinking, his leasing scheme was a
master stroke. Furthermore, since he was living in Toronto he did not have to
face the issues and pressures associated with living in the midst of settlers who
often attacked the company for any ills in the Huron Tract, perceived or real.

Jones' House in Goderich. Following Thomas Mercer Jones' release from Canada Company employ in 1852, he and Elizabeth Mary moved across the road to this house on West Street. He became agent for the Bank of Montreal in Goderich from 1853 to 1857. When his wife, who had been in ill health, died, Jones returned to Toronto. He died there in 1868 "from the effects of drink." *Courtesy of Paul Carroll.*

Widder was smart enough to not interfere while Jones was being drawn into the politics of the day without his really comprehending what was happening to him. Widder was a "big picture" person who could deliver – and did. The directors liked that. At the same time, they do seem to have sensed the difficult position which Jones was in and were willing, to a point, to overlook his weak points in handling the business of company in the Tract. Furthermore, they could not be sure that another person would do any better. Jones' marriage to Bishop Strachan's daughter had a downside, of course, in that as Jones' judgment began to slip, the directors would have known that if they decided to dismiss Jones sooner, they could expect a vitriolic and public attack by Bishop Strachan. The company did not want this, particularly after the Commission of Enquiry and the election of 1841. While Strachan's blistering, negative attack as "Aliquis" had been a positive for the company in the end, they did not want to risk another public skirmish. They were willing "to leave well enough alone."

In retrospect, Jones' day-to-day work in the Huron Tract generally satisfied the directors and his Strachan connection was seen as a "plus." As to Widder, he was increasingly looked at to play a creative role in Toronto – which he did with great success.

As long as the two commissioners worked reasonably well together, there was every reason to maintain the status quo. However, when the growing animosity between the two men surfaced, Jones was seen as the culprit. Furthermore, as errors in company accounts from Goderich became more prevalent, Jones was increasingly viewed as not being up to the job. He could not, or would not, understand this and wrote to Charles Franks in February 1853 to complain of his treatment in the hands of the directors. Franks replied:

> The step they [the Directors] have had to take in dispensing with your services as a Commissioner, has been I assure you a matter of pain, and very sincere regret to them. In regard to the cause, I cannot explain it better than I did in my letter of the 18th Novr. last in which you will find the following passage "you will recollect when you were in England some few years since how anxiously I endeavoured to impress upon you the importance of perfect harmony both in opinion and conduct between those who represented the Company in Canada in which I think you certainly have not succeeded." The Directors were convinced this harmony did not exist, that it was their duty to take such steps as would ensure it and the course they adopted in altering the management of the Company, was the necessary consequence.[155]

The dismissal of Thomas Mercer Jones was clearly the result of the Canada Company's man in the Huron Tract operating increasingly as an independent entity. Furthermore, he had developed "localitis" and his inattention to detail was impacting negatively on the company. If any good came from the railway issue it was that the directors had the "hook" they needed to terminate him. Jones had precipitated his own downfall by becoming as stubborn as John Galt.

CHAPTER 6

YOU BE THE JUDGE

How useful it [the Canada Company] was to
Upper Canada is ... another matter.[1]

O PINIONS ARE AS VARIED as those who form them as to the particular
amount of credit or discredit, earned or otherwise, with which the
Canada Company had to live during the 1826-1853 time frame. To be objective
in forming an opinion of the company and its operations in Canada, one must
think in terms of the first half of the nineteenth century. Britain faced a heavy
drain of capital as a result of the Napoleonic Wars and was forced to re-align
her policies in North America. In looking for an answer to her financial problems regarding Upper Canada, the plan proposed by Robert Gourlay, John Galt
and John Beverley Robinson for the orderly settlement of the colony, coupled
with a guaranteed income, seemed to offer the best solution. Undoubtedly, had
the British government not agreed to sell that vast acreage to the Canada
Company, a great deal more money would have accrued to the provincial treasury in later years. However, the thought of a guaranteed sum for uninhabited
forest land in the early years and a more orderly settlement of the Crown
Reserves, along with promotion of, and settlement in, the recently acquired
lands beside Lake Huron was just too attractive to turn down.

By 1853 John Beverley Robinson had been knighted and was now chief
justice of Canada West. He observed that year that had the government held on

The falls on the Maitland River near Goderich was one of the best sites
for a grist and lumber mill. Note the quality of construction of the mills.
Courtesy of the Archives of Ontario F 129, ACC 1372 AO 6477.

to the lands sold to the company and managed them by adding two or three
clerks to the land offices, almost 20d per acre would have been realized by that
time. He lamented:

> What the interest of such a fund would have enabled the
> Government to do for the Province in promoting education
> public improvements ... defences etc. can hardly be expressed.[2]

But he also noted that there would have been little hope for procuring and dis-
pensing such benefits through the agency of the government because "the
Assembly would have clutched all."[3]

In reaching any conclusion pro or con on the Canada Company and its operations in Canada, it is important to remember that their lands could not be
priced above the going rate because of competition from sheriff's sales and
other government, clergy, school and college lands which were also on the
market. Therefore, in order for the company to remain financially sound and
attract/maintain shareholder support, there had to be rigid control over the
company's finances. As far as the directors were concerned, if savings could be
achieved through the holding back of company expenditures, there was every

reason to do so. Furthermore, if the policy of not undertaking projects which could be run by private enterprise seemed questionable, the first gristmill built at Goderich underlines the problem faced by the company. Unquestionably, a gristmill is one of the first requirements of any settlement, yet there was scarcely enough grain handled at the Goderich mill to keep it going two days a week. In the final analysis, if company policy seemed one-sided and short-sighted, the British and colonial governments do not escape blame. After John Galt's departure and William Allan's arrival, the company became very well-connected with the Family Compact. Despite the condition that all expenditures out of the Huron Tract Improvement Fund required prior approval by the Executive Council, few, if any, expenditures were subjected to scrutiny beforehand by the Council – and most final costs came in well above estimates, with nary a complaint. Furthermore, because the £48,000 allocated for the Huron Tract Improvement Fund had to last until 1843, the company had to be very judicious in the use of the monies in any given year. Managing the needs of the colony with the realities of the fund became a constant balancing act.

John Galt's Tomb in Greenock. John Galt died in 1839, wracked with pain.
He never recovered financially after returning to Britain. Following his death,
his wife Elizabeth returned to Canada and lived with her son Alexander Tilloch Galt
in Montreal who went on to become a "Father of Confederation." Note the flowers
from the Town of Goderich in this photo, taken in 1927 on the 100th anniversary
of the founding of the town. *Courtesy of the Archival and Special Collections,*
McLaughlin Library, University of Guelph, H.B. Timothy Collection.

The Canada Company was a public stock company and was expected to make a profit for its shareholders. While the directors had a commitment to the government vis-à-vis fulfilling their contract, they were also obliged to retain shareholder support. Their declared objective was not unreasonable, i.e., to promote and accelerate immigration to the province and to increase the value of their lands with a view to profitable resale. Such a policy required commissioners in Canada who were prepared to run company affairs from such profit-oriented motives, while at the same time remaining on favourable terms with both the colonial government and the settlers. The company directors understood this from the beginning, but the task was not going to be easy. Indeed, John Richardson, the influential Canadian businessman who had been appointed a referee to resolve the Clergy Reserves issue in 1825 wrote to John Strachan in 1824:

> The Canada Company will produce most beneficial effects if
> well managed, but the great difficulty will be to get men of
> integrity and talent to execute the details.4

Scottish-born John Galt, the Canada Company's first commissioner in Canada, had an independent streak and invariably would not follow directives from London. An idealist, he refused to believe that the company could not afford to carry on as he had visualized. Furthermore, given his behaviour, the relationship between the company and the colonial government had hit an all-time low by the time he was released by company directors early in 1829. Because of his attitude and work habits, including his inexplicable lack of attention to detail, the directors had little choice but to terminate him.

When William Allan and Thomas Mercer Jones took over from Galt, the operations in Canada were in a state of interregnum. The new co-commissioners were appointed on a temporary basis only and many in the know wondered whether the company would survive. Control at this point was firmly held in London. It was only relinquished when shareholders signified their continued support of the company, and the capabilities of the new commissioners had been demonstrated. Allan and Jones worked closely with the directors in England. A solid relationship developed. The 1830s, though, were a trying period due to the cholera epidemic, and the unsettled political situation in Upper Canada which lead to the Rebellion of 1837, but generally speaking, the

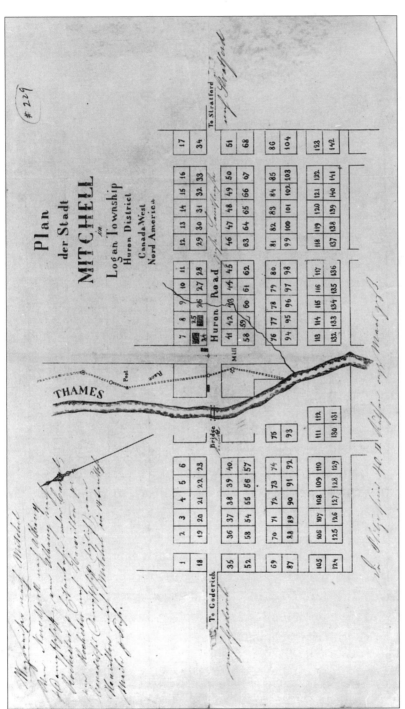

The site for Mitchell, at the junction of the Thames River and the Huron Road was originally called "Big Thames" by Tiger Dunlop. Laid out by the Canada Company on 400 acres of land, the first lots were sold in 1836. The company actively promoted land sales in Germany, hence the description "Plan der Stadt Mitchell." It became a village in 1857 and was incorporated as a town in 1874.

Courtesy of the Archives of Ontario, F129 TRHT H 229 AO 6486.

Bank of Upper Canada, Goderich. This building on Coburg Street was constructed
circa 1853 as the bank premises and residence for the manager of the bank.
It subsequently became the Presbyterian Church manse. *Courtesy of Bruce Sully.*

directors were pleased with their commissioners in this difficult period. Jones,
however, tended to be erratic and not always successful in his endeavours, but
his marriage to Bishop Strachan's daughter, Elizabeth Mary, apparently made
up for his shortcomings, at least during the earlier years.

The arrival of Frederick Widder and the retirement of William Allan almost
two years later changed company practices in Canada considerably. Over a
period of time, Widder's drive, enthusiasm and approach to business stole the
spotlight from Jones who had been sent to Goderich to manage the Huron
Tract lands. While Jones' actions and attitudes had remained "provincial,"
Widder's horizons were much broader. He pursued new approaches to marketing company lands and was particularly effective in his efforts to increase sales
by way of his innovative leasing scheme. From then on, he was regarded, at least
unofficially, as the senior commissioner. Not since its inception had the
company received such favourable publicity as that generated by Widder. The
Kingston Chronical and the *British Colonist* of April, 1845 reported:

> ... before that gentleman took charge of the Company's affairs,
> there was not a more unpopular body in the Province, either in

This 1867 photo shows Goderich Harbour during the spring breakup. The ships in port include the two gunships sent the previous year to protect Goderich and the harbour should Fenians decide to attack. Note the Buffalo and Lake Huron Railway grain elevator which had just been built. *Courtesy of the Huron County Museum, Neg. x985.0049.096.*

Parliament or out of it – and that since Mr. Widder has introduced his more liberal and comprehensive system – independent of political and party views – he has lived down formerly existing prejudices; and we speak advisedly when we say, that there is no public body now more fully appreciated, or one in whom more confidence is placed, by all classes of society and by every party.[5]

Towards the end of Jones' tenure, tensions between the two commissioners had escalated and when Jones sided with the Huronites over the railway issue in opposition to company policy, that was more than the directors could tolerate. Having become increasingly uneasy with him, they had started to build a case of grievances. The directors now had the excuse they needed to terminate him.

The new and effective Widder/Robinson duo would run the company's operations solely from Toronto. Local politics and the company's books relating to the Huron Tract would now be under control, at least from the company's perspective.

In the final analysis, the directors had been willing to allow their commissioners in Canada as much freedom as they thought commensurate with circumstances. Galt had been given considerable scope, but had misused his power and authority. Furthermore, his work habits were sloppy. Allan and Widder,

By the 1870s, Goderich was developing into a busy port. In 1872, the Maitland River estuary was to change with the channeling of the river and the beginnings of today's harbour. That same year, Goderich was named a Harbour of Refuge which would have meant federal money for improvements. *Courtesy of Mac Campbell.*

both men of executive ability, performed well. Jones, on the other hand, while known to be hard-working and faithful to the company until close to the end, had neither the business acumen, judgment, nor the diplomatic skills of Allan and Widder, and was prone to forgetting his limits.

The larger-than-life personalities of the day and the requirement that the company operate from the perspective of a shareholder-driven, profit-making enterprise in a challenging political environment had clearly made for exciting times during the company's first 27 years.

EPILOGUE

T HAT THE CANADA COMPANY was a force to be reckoned with in the early days of Upper Canada/Canada West is very obvious and well-known to those Canadian historians who focus on Ontario history. That it continued to exist until 1953 is not so well-known, because after the tax dispute of the 1840s, the company was able to settle down to perform its stated function as a seller of land utilizing its highly successful leasing program. Furthermore, the press it was now receiving was generally favourable. As such, it no longer figured prominently in the politics of Ontario, although it did continue for a number of years to have at least one commissioner politically "well-connected" in the province.

By 1856 the company found itself in an almost embarrassing financial position. While significant sums had been spent on purchasing and improving their lands, a large portion of their property had been sold at prices exceeding these costs by quite a considerable margin. Furthermore, the value of unsold lands far exceeded the paid-up capital of £298,737/10 and other liabilities of the company. Accordingly, in June 1856, by British Act of Parliament, the directors declared that "it should not be lawful ... to call up any further part of the subscribed capital of the Company."[1] The act further stated that it would be lawful for the proprietors at a special general court to direct that the affairs of the company be wound up, i.e., that the company should be dissolved as soon as the liabilities and debts had been paid and satisfied and all assets realized and divided amongst the shareholders.

Goderich Harbour circa 1860. The Canada Company constructed the
first piers in Goderich with the launching, in 1831, of the *Lady Goderich* built to
bring in settlers from Detroit. Given the condition of the roads, the port was
particularly vital to the town's economy prior to the arrival of the railway in 1858.
The author's great-great grandfather, William Lee, and his grandfather, Charles C. Lee,
were ship chandlers there. *Courtesy of Mac Campbell.*

That said, the company was to continue for another ninety-seven years. By
1878 the sum of £31/10 per share had been paid back to the shareholders,
leaving £1 per share or £8,915 in capital stock to be repaid.[2] The directors now
saw their purpose as leasing and selling lands, collecting rents and royalties for
minerals, selling timber and winding up the business of the company.[3] In 1916,
however, the directors decided to enlarge the scope of the company. The
"Canada Company's Act" of that year stayed the winding-up of the company,
and enabled it to allot and issue further share capital of £100,000 for existing
shareholders (the actual share capital issued was £89,190)[4]. Quite considerable
amounts of company land in the Huron Tract (29,933 acres), Crown Reserves
(59,270 acres) and park lots (268 acres) still remained unsold[5] and the directors
felt it would take some time before final dissolution would be possible. Because
of this and, in order to take part in the industrial boom resulting from the First
World War (1914-18), the objects and powers of the Company were extended
to enable it to, in part:

> a) purchase ... hold, sell, exchange lands, water lots, mines,
> metals, minerals, mineral oil, natural gas and quarries ... in the
> Dominion of Canada and the Colony of Newfoundland.

Market Square, Stratford, circa 1880. Stratford became the county seat in 1853 with the separation of Perth County from the Huron District. It also became a major rail centre with the arrival of the Grand Trunk Railway and then the Buffalo and Lake Huron Railway in 1856. The building in the background centre is the Commercial Hotel which was owned and operated by Timothy Hagarty. *Courtesy of the Stratford-Perth Archives.*

b) … work and stock and make merchantable and sell related produce.

c) purchase … construct equip work maintain improve alter … any railway, tramways, engineering works, electrical works, freezing works, gas and coke works, abattoirs, wharves, docks, bridges, roads, buildings or erections ….[6]

Dividends were to be paid from any profits, and the par value per share was to be £1/0/0.[7]

Despite the act, little appears to have been done with the enlarged powers. In 1938, the company now held a total of 23,711 acres of unsold land[8] and the par value per share of company stock was reduced to ten shillings.[9] In March 1948, the company no longer saw itself as "a valid and subsisting corporation, leasing and selling lands, and collecting rents and royalties for minerals."[10] The

Goderich Harbour, circa 1900, as shown in a photograph by Reuben R. Sallows[12] of Goderich. This view of the harbour shows the degree of commercial activity there by now. There were boat builders, hotels, taverns, fisheries, warehouses, a soap factory, a grain elevator and sheds and a flour mill. *Courtesy of the Sallows Gallery & the Huron County Museum.*

following year, it became a "Land Company in process of liquidation"[11] and, in March 1950, the company listed its holdings as 4,207 acres in the townships of Bosanquet (Lambton County), Lansdowne (United Counties of Leeds and Grenville) and Ramsay (Lanark County).[13]

An offer from the Province of Ontario to buy the Canada Company's remaining lands in Bosanquet Township for the establishment of the Pinery Provincial Park provided the catalyst for the directors to start the process of winding up. This area was poor farmland and had already been repossessed once. By the end of 1951, the Pinery lands and all other remaining company property had been sold. On June 26, 1953, the company was registered under the English Companies Act for the purpose of liquidation. A "members' voluntary winding-up" began on August 12, 1953,[14] and the liquidation process was completed on December 14, 1953.[15] Interestingly, the Extra-Provincial License for the Company to operate in Ontario was only cancelled on September 11, 1961.[16]

At the 127th Annual Canada Company Court in London, England, on January 21, 1953, the company was dissolved. The end of an era had been reached. Upper Canada had changed from an untamed wilderness to the modern province of Ontario, and the Canada Company had outlived its purpose.

APPENDIX A

CHRONOLOGY OF EVENTS

1791:   Constitutional Act creates provinces of Upper and Lower Canada. Throughout each newly created township in Upper Canada, 200 acre Clergy and Crown reserves are set aside representing two-sevenths of each township.

1812:   U.S.A. declares war on Great Britain at height of Napoleonic conflict and strikes at Canada.

1815-1824: Upper Canada faces financial hardship due to war and its aftermath, including property damage claims and military pensions. British Corn Laws inhibit exports of wheat and flour into British market. Land tax not judged possible.

1822:   Robert Gourlay proposes sale of Crown Reserves to pay for war losses in his *Statistical Account of Upper Canada*.

1823:   John Beverley Robinson, attorney general of Upper Canada, appears to be first person with credibility to suggest selling Crown Reserves to a company to raise funds to apply against day-to-day costs of running the province, including war reparations/pensions. Clergy Reserves are then added to mix. John Galt would argue he was the first to promote the concept.

This sketch shows the Priory and Allan's mill, circa 1853. Guelph now has additional buildings. In 1846, Smith's *Canadian Gazetteer* reported that the village had a population of 1,240 persons while the county had 3,400 residents. Allan's mill was supplying the needs of these settlers. *Courtesy of Archival and Special Collections, McLaughlin Library, University of Guelph, Allan-Higinbotham Papers.*

1824   Canada Company founded in London, England, with Provisional Committee created to run it until charter granted. Capital set at £1,000,000 sterling (10,000 shares at par value of £100 each). Shareholders are assessed an initial £5.

1825:   Crown purchases land covered in part by the Huron Tract from the Chippewa First Nation for £1,100 – to be paid annually.

1826:   Canada Company obtains charter (August 19).

Reaches agreement to purchase 1,000,000 acres in Huron Tract in lieu of Clery Reserves (829,430 acres), plus Crown Reserves of 1,384,413 acres and 42,000 acres Halton Block at a cost per acre of 3s/2d sterling or 3s/6d currency. John Galt is first commissioner.

1827:   April 23: Galt founds Guelph without name approval from directors.

June 29: Galt, with Tiger Dunlop, founds Goderich.

1826-1829: Galt annoys company directors and colonial authorities.

1828:   Company shares fall to £6, a 40% discount over paid-in value. Company receives additional 100,000 acres in lieu of land unfit for cultivation in Huron Tract.

Commemoration of Shakespeare's Centenary, Stratford. On April 23, 1864,
an oak tree was planted in Shakespeare Place (now Memorial Park) to mark
Shakespeare's centenary. Little did Canada Company commissioner
Thomas Mercer Jones know the impact of his naming the community
"Stratford," and the river, the "Avon." *Courtesy of the Stratford-Perth Archives.*

1829:   Two new company commissioners announced at company board
         meeting (Thomas Mercer Jones and William Allan); Galt receives
         recall/termination letter.

         Company per-share value falls to £1/10/0 from paid-in value of
         £12/10.

         Allan opines: lack of shareholder support result of misunderstand-
         ing of nature of investment.

1830:   Shareholders approve continuation of company. Extensive advertis-
         ing campaign initiated.

1831:   Overland stage route established between Hamilton and Goderich
         and *Lady Goderich* launched to bring in settlers to Huron Tract by
         water.

Baron de Tuyll buys 4,000 acres from company for projected city, Bayfield, at mouth of Bayfield River south of Goderich.

1832:   Jones names village at east end of the Huron Road "Stratford" and river, the "Avon."

1833:   Company share value recovers.

1832–1835: Popular press criticizes company

1835:   Huron Union Society demands responsible government, sale of Clergy and Crown reserves by government and abolition of all licensed monopolies, most notably, Canada Company. Government of Upper Canada's *Seventh Report from Committee on Grievances* attacks company.

1839:   Frederick Widder appointed Canada Company commissioner.

Canada Company office in Goderich opens with Jones in charge.

Stage line from London to Goderich established.

Legislature of Upper Canada launches Commission of Enquiry into affairs of Canada Company. Comes to naught.

1840:   William Allan leaves company.

1841:   Canada East and Canada West created with single government and legislature.

District of Huron created for legislative and land purposes. It includes all of the townships of the Huron Tract (except for Bosanquet, now part of Lambton County) and Crown Lands one range to the north which are separated from the London District to become part of the Huron District.

1842:   Leasing plan proposed by Frederick Widder adopted in Huron Tract.

The Canada Company probably sold the Priory to William Allan in 1832
when he purchased the Canada Company gristmill which was close by.
Allan is pictured here with his wife, their two children and gardener.
The family lived here until his death in 1857 when his son David Allan took
over both the home and the mill. *Courtesy of the Guelph Public Library Archives.*

1842:   Jones chastised for laxness at operation in Goderich. Widder praised.

1843:   Leasing plan great success; extended to all company lands.

1849:   Goderich incorporated as town.

1850:   Huron, Perth and Bruce become separate counties, but are known
          as United Counties and are administered as unit.

1852:   Jones' employment with Canada Company is terminated. Widder is
          now chief Canada Company commissioner. William Benjamin
          Robinson replaces Jones. Goderich office reduced to agency status.

          Stratford becomes village.

Railway Station, Guelph, circa 1900. The Priory became a railway station
in 1887 when the Guelph Junction Railway tracks were laid along the west bank
of the Speed River through the property's former garden and orchard. In 1911 a new
station was built elsewhere and the building was shifted to a vacant lot. Uncared for,
it caught fire in 1926 and suffered damage to the roof. The building was then
condemned and torn down. *Courtesy of the City of Guelph Archives.*

Perth withdraws from United Counties and becomes a separate county.

1878:   By this date, shareholders have recouped £31/10, leaving £1 per share, or £8,915 in capital stock, to be repaid.

1953:   January 21: Directors hold 127th Annual (and last) Canada Company Court (or Board of Directors meeting) in London, England. Company is liquidated by year end.

APPENDIX B

THE MINUTES OF JULY 30, 1824 record the first Canada Company Court (Board) of Directors' meeting. At this meeting, it was unanimously resolved that the following would hold positions on the Provisional Committee of the company:

Directors: Robert Biddulph, Richard Blanshard, Robert Downie, John Easthope, Edward Ellice, John Fullarton, Charles David Gordon, William Hibbert, John Hodgson, John Hullett, Hart Logan, Simon McGillivray, James McKillop, John Mastermen, Martin Tucker Smith, Henry Usborne and William Williams.

Charles Bosanquet was elected chairman (or governor) and William Williams as deputy chairman (or governor). The auditors were to be Thomas Harling Benson; Thomas Poynder, junior; Thomas Wilson and John Woolley.

John Galt was appointed secretary and Messrs. Freshfield and Kaye were chosen as company solicitors. The bankers were to be Messrs. Masterman and Company and Messrs. Cocks, Cocks, Ridge and Biddulph.

Having received its charter on August 19, 1826, the Canada Company's Court of Directors met on August 24 1826 to put a structure formally in place. The first order of business was for John Galt as secretary to table the abstract of the charter. It declared, inter alia, that the proprietors (shareholders) would be "one body politic and corporate in deed and in name by the name of "The

John Galt, (circa 1834), began his career with the Canada Company as secretary. Following his termination in 1829 and his return to London, Galt spent five months in debtors' prison for unpaid tutor's bills for his sons. While there he wrote prodigiously. Along with former Canada Company director, Edward Ellice, he promoted the idea of a land company in Quebec's Eastern Townships (The British American Land Company). He was company secretary for ten months in 1832, but ill health forced his resignation. *Courtesy of Archival and Special Collections, McLaughlin Library, University of Guelph, H.B. Timothy Collection.*

Canada Company." The capital of the company was not to exceed one million pounds on shares of £100 each. Company proprietors or shareholders who held a minimum of 25 shares were eligible to become directors.

The company directors and secretary were formally declared to be those who had served on the Provisional Committee and who were listed in the company's charter. The bankers and auditors were confirmed as those who had been appointed two years earlier. James William Freshfield of Messrs. Freshfield and Kaye was named solicitor.

The directors and auditors listed in the charter document were to stand until March of 1829 when one third were to be rotated off the Court of Directors each year by drawing lots (by that time, the number of directors had actually dropped to twelve from eighteen due to resignations). Directors were eligible for re-election. The governor and deputy governor were to remain in office until 1831.

The Minutes of the Court of Directors of February 19, 1829 record the system of drawing lots: Slips of paper numbered one to twelve were to be placed in a "balloting glass." The directors were then to be called in alphabetical order by surname, with the exception of the chairman of the Committee of

Correspondence who was previously determined to be number 12.

Following these instructions, the directors drew one slip each from the bowl while the chairman drew a slip for each of the absent directors. The results in order were a follows:

> To leave office in March 31, 1829: John Easthope; John Hullett; Henry Usborne; Edward Ellice;
> To leave office in March 1830: Hart Logan; James McKillop; William Hibbert; John Fullarton;
> To leave office in March 1831: Robert Biddulph; Robert Downie; Martin Tucker Smith; Simon McGillivray.

Although Edward Ellice was to leave office in March 1829, the directors recommended to the proprietors that he be appointed deputy governor in place of William Williams who had resigned. They concurred.

To replace those directors who had left the Court, printed slips were prepared containing the names of all shareholders who were entitled to vote. The voting procedure was as follows: holders of between five and less than ten shares had one vote; holders of ten and less than twenty shares had two votes; holders of twenty and less than twenty-five shares had three votes; holders of more than twenty-five shares had four votes.[1]

APPENDIX C

HURON TRACT TOWNSHIP NAMES AND THEIR ORIGINS

CANADA COMPANY DIRECTORS agreed to the creation of twenty-one townships in the Huron Tract and named five of them after senior members of the British government, namely Colborne, Goderich, Hay, Stanley and Stephen. The balance, or sixteen townships, were each named after company directors of which there were eighteen when the company received its charter. The exception was Easthope Township which was split into North and South Easthope (probably because the tri-angular piece of land would be too large if not split, but too small to be split and named separately for a director). How it was decided who would, or would not, be immortalized is not clear.

All of the Canada Company directors were resident in England when the company was formed, but at least four had lived at various times in Lower Canada namely, Edward Ellice, Simon McGillivray, Hart Logan and Henry Usborne. This group knew the political/economic situation in the Canadas well.

Scottish-born Deputy Provincial Surveyor John McDonald, on behalf of the company, made all of the original surveys of the townships, or they were completed under his direction. The exception is Goderich Township. This particular survey was completed by the Deputy Provincial Surveyor David Gibson. McDonald also surveyed the road from Guelph to Goderich which was cut through at the expense of the Canada Company. He was assisted by Deputy Provincial Surveyor David Smith and McDonald's cousin Donald McDonald.

Spring Creek Mills, Goderich Township. Thomas Trick purchased 79.25 acres in 1873. Originally acquired by an Alexander Grant from the Canada Company prior to 1854, the property became known as "Spring Creek Mills." The farm is still owned by the Trick family. *Courtesy of the Huron County Museum (Toronto: H.R. Beldon & Co. 1879), Gorrell & Co. Toronto Litho Co.).*

Trick Mill, Goderich Township with mill on left and dam in background. Farmers came from miles around to have their grain ground and lumber cut at this mill. To-day, the gristmill building has a new roof, but it still has the original siding. It currently produces electric power for the Trick farm from an 1860/70s turbine. *Courtesy of William Trick.*

Sir John Colborne (later Lord Seaton) was arguably the ablest lieutenant governor of Upper Canada (1828–1837). Unwilling to support Strachan's proposal for a university, he established a preparatory school for boys, Upper Canada College, with public school/university monies. Colborne Township north of Goderich was named in his honour. *Courtesy of the Upper Canada College Archives.*

Biddulph Township

Robert Biddulph: member of Canada Company's Provisional Committee formed in 1824; company director; partner in firm "Messrs. Cocks, Cocks, Ridge and Biddulph;" Canada Company bankers; rotated off the Court of Directors in 1831.

Blanshard Township

Richard Blanshard: member of Canada Company's Provisional Committee formed in 1824; followed Charles Bosanquet as governor of company but had resigned by February 1829.

Bosanquet Township

Charles Bosanquet, MP: chairman of Canada Company's Provisional Committee formed in 1824; first governor (or chairman) of the company; member of the Committee of Correspondence; had resigned by February 1829.

Colborne Township

Sir John Colborne: distinguished military career; reported "to have fought brilliantly" at the Battle of Waterloo; lieutenant governor of Guernsey (1821–28); promoted to major general in 1826; in general terms, arguably Upper Canada's most able lieutenant governor (August 1828-January 1836) – among other things, he vigorously promoted public works and promoted immigration. By favouring British immigrants, however, and by diverting funds in unpopular ways including putting public school funds and money earmarked for the establishment of a university into the creation of the classical English preparatory school, Upper Canada College, Toronto in 1829, he created unrest. This contributed to the Reform victory in 1834 and to the rebellion in Upper Canada in 1837; after his recall as lieutenant governor in 1836, Colborne was subsequently placed in command of British forces in Canada; he and his troops suppressed the insurrection in Lower Canada in 1837 and crushed the second rebellion there in 1838; he was briefly governor general of Canada in 1838/39. In October, 1839, Colborne returned to Britain and was elevated to the House of Lords as Lord Seaton of Seaton; spoke against the Act uniting the Canadas; Canada Company directors had initially wanted this township called "Horton" after the influential Undersecretary of State for Colonial Affairs, Robert Wilmot-Horton. There already was a Horton Township near Ottawa.

Downie Township
Robert Downie, MP: member of Canada Company's Provisional Committee formed in 1824; company director and member: Committee of Correspondence; rotated off Court of Directors in 1831.

Easthope Townships (North and South)
Sir John Easthope, MP: politician, journalist, stockbroker; member of the Canada Company's Provisional Committee established in 1824; company director; magistrate for Middlesex and Surrey; chair of London and South-Western Railway Company; chair of Mexican Mining Co.; business partner of Simon McGillivray, including ownership of *Morning Chronicle and London Advertizer* newspaper (purchased in 1835); rotated off the Court of Directors in 1829.

Ellice Township
Edward "Bear" Ellice, MP: heir to substantial fortune; highly successful businessman; power behind amalgamation in 1821 of XY Company/North West Company (NWC) and Hudson's Bay Company (HBC); deputy governor of HBC; member of Canada Company's Committee of Correspondence; as a politician in the 1820s, was principle spokesman for the Whigs on economic questions; in

early 1830s, became secretary of the Treasury and chief whip; in 1833, entered the British cabinet as secretary of war; left politics in 1834 to manage business interests. Although he was to be rotated off the Canada Company's Court of Directors in 1829, was the unanimous choice to replace William Williams as deputy governor (Williams had unexpectedly resigned the previous month). Ellice was the largest absentee landlord in Lower Canada. With John Galt in the early 1830s, he successfully promoted the concept of establishing the British American Land Company along the lines of the Canada Company. It was incorporated in 1834 and acquired 850,000 acres (344,000 ha) of Crown Land in the Eastern Townships of Quebec with headquarters in Sherbrooke, Quebec.

Fullarton Township
John Fullarton: member of Canada Company's Provisional Committee established in 1824; company director and member, Committee of Correspondence; rotated off the Court of Directors in 1830.

Goderich Township and Town of Goderich
Right Honourable Frederick John Robinson, MP: first elected to the House of Commons in 1806; held wide variety of very senior positions in government, including president of the board of trade (1818-1823; 1841-43); in 1826, as chancellor of the exchequer and member of the House of Commons supported legislation to incorporate the Canada Company; created Viscount Goderich of Nocton on April 28, 1827; appointed secretary of state for war and colonies in April 30, 1827; was subsequently appointed to House of Lords and became government leader there; appointed prime minister (August, 31,1827 to January 8, 1828 when he tendered his resignation); November 22, 1830 – appointed secretary of state for war and the colonies – resigned in early 1833; April 13, 1833 – created Earl of Ripon. While Lord Goderich was an "amiable man," he was seen as a very weak prime minister who found stress difficult to handle (see Chapter 2, Note #33 for a more fulsome account of Lord Goderich).

Hay Township
Robert William Hay: second undersecretary of state for colonies (1825); became first permanent undersecretary responsible for overseeing the work of the North American department for most of the period 1828 to 1836.

Hibbert Township
William T. Hibbert: member of Canada Company's Provisional Committee established in 1824; company director; rotated off the Court of Directors in 1830.

Hullett Township
John Hullett: member of the Canada Company's Provisional Committee established in 1824; company director; partner in banking firm of Messrs. Hullett Brothers and Company; rotated off the Court of Directors in 1829. John Galt was asked early in 1824 by Colonial Undersecretary Robert Wilmot-Horton on behalf of Colonial Secretary Lord Bathurst if he could find purchasers for the Crown and Clergy reserves in Upper Canada. Galt contacted Hullett's firm which responded in the affirmative.

Logan Township
Hart Logan: member of Canada Company's Provisional Committee established in 1824; company director; had long business association with Canada (including principal in import house founded in 1796 in Montreal); shipowner; builder of sailing ships and steamships; partner in export/import business based in Liverpool and London trading into Europe, Canada and the Caribbean; "Messrs. Hart Logan" was Canada Company agent in Montreal; rotated off the Court of Directors in 1830; became member of the Committee of Correspondence of the Lower Canada Land Company's London subscribers (company was precursor of the British American Land Company successfully promoted by Ellice and Galt in early 1830s).

McGillivray Township
Simon McGillivray: businessman and leading partner in North West Company (NWC); in 1821, with Edward Ellice, devised solution to merge XY Company/NWC and Hudson's Bay Company (HBC) into HBC; member: joint HBC/NWC board to manage fur trade according to the new configuration until board was dissolved in 1824; member, Canada Company's Provisional Committee established in 1824; was to rotate out of office in March 1831 but resigned in September, 1830 over his remuneration as chairman of the Committee of Correspondence; had a number of other ongoing business interests. At age 54, married Anne Easthope, daughter of business partner and fellow Canada Company director, John Easthope; provincial Grand Master of Masonic movement in Upper

Canada for 18 years. In 1829, helped organize United Mexican Mining Association of London; in 1835, with John Easthope, bought *Morning Chronicle and London Advertizer* newspaper.

McKillop Township

James McKillop, MP: member of Canada Company's Provisional Committee established in 1824; company director; became deputy governor of company; rotated off the Court of Directors in 1830.

Stanley Township

Edward George Geoffrey Smith Stanley, 14th Earl of Derby, MP: Latin scholar, skilled debater; distinguished parliamentary undersecretary of state for the colonies (1827-28) including when Lord Goderich was prime minister; secretary of state for the colonies (1841-45); member of parliament: 1822 to 1844 when he was elevated to the House of Lords; named chancellor of Oxford University in 1852; was called upon three times to form government between 1851 and 1855; prime minister twice for very short periods of time; was shrewd in business and passionate about sports.

Stephen Township

Sir James Stephen: legal advisor to the colonial office in 1820s; was then second undersecretary for the colonies to 1836; permanent undersecretary for the colonies: 1836 to 1847; played important role in shaping British government's North American policy.

Tuckersmith Township

Martin Tucker Smith (surname was sometimes hyphenated): banker and member of Canada Company's Provisional Committee established in 1824; company director; referred to as Martin T. Smith on Canada Company posters; rotated off the Court of Directors in 1831.

Usborne Township

Henry Usborne: member of Canada Company's Provisional Committee established in 1824; company director; in 1801 started successful timber and sawmilling operations in Lower Canada; owned 45,000 acre seigneury in the

Gaspé region; also owned timber-related business in Britain; rotated off the Court of Directors in 1830.

Williams Township

William Williams, MP: deputy chairman of Canada Company's Provisional Committee established in 1824 and subsequent Court of Directors; resigned in February 1829. "I have long felt uncomfortable at retaining my situation … without being able to attend any of the duties" (as deputy chairman). He added that he did not like the fact that there were no emoluments attached to the position.

APPENDIX D

THE HURON TRACT

D R. WILLIAM "TIGER" DUNLOP would have written this piece after he left the Canada Company in 1838. It was published in 1841, the year his brother, Captain Robert Graham Dunlop, died. "Tiger" was now preparing to run for election in place of his brother who had served in the Assembly of Upper Canada from 1836 to 1841. The election of 1841 was to elect a representative to send to the new legislature of the united Canadas.

The Huron Tract described here is larger in area than the original Canada Company lands.

The Huron Tract

The county of Huron is of a triangular shape; its base, which is 70 miles in length, resting on Lake Huron. To the north of it is a tract of government land, now being surveyed, which will be added to the county. The part at present surveyed is about 1,200,000 acres, and where the whole is surveyed, it will amount to about 1,700,000 acres.

The whole of this tract is watered in every direction by the Thames, the Sables, the Bayfield, the Maitland, and a considerable river in Ashfield. In the eastern part of the county the Nith river, and in the west are numberless streams of various sizes falling into Lake Huron.

The land generally, is of a loamy description; sandy loam with limestone gravel on the verge of the lakes, and clay loam towards the interior, and everywhere covered with a considerable depth of vegetable mold; and the whole county may he said to be bedded at various depths on a recent limestone formation, though sometimes this is varied with sandstone, which however is not pure, but seems to have been a stratum of sand bound together by lime. At Kettle Point, there is a formation of clay slate, having embedded in it globular pyrites. On the lake shore there are found detached masses of serpentine, and in every part of the tract, masses of red, silver and gray granite.

The whole of the land is of excellent quality. There is an extensive cedar swamp, which commences in the township of Ellice, and running through Logan, Mackillop [and] terminates in Hullett. This, to be made available, would require to be drained, but that would be no difficult matter, as it is the summit level of the whole country, and from the springs in this swamp arise many of the rivers which fall into Lakes Erie, St. Clair, and Huron. Once drained it will be the richest land in the county. The only really bad land in the Tract, is a narrow strip of sand of eleven miles in length, and of hay a mile to a mile in breadth, which lies between Lake Huron and the River Sables, near its mouth.

The principal timber is maple, elm, beech and bass, and in lesser quantity cherry, hickory, ash, oak, hemlock, and pine, the latter however being very scarce. Black Walnut grows in the South part of the tract. The rivers and lake abound in fish, among which may be enumerated the sturgeon, river trout, pickerel, pike, muskellunge, mullet, carpe etc. The game common to the provinces is abundant, except in the townships of Goderich and Colborne.

There is abundance of magnetic iron-sand on the shores of the lake, and some slightly salt and chalybeate springs in different parts of the county. There is a good deal of sulphur at Kettle Point, and lead ore exists in the rear of it.

The whole country is extremely healthy, intermittent and resistant fevers, are unknown, except when imported. This arises from the whole county being a table land, varying from 120 to 300 feet above the level of Lake Huron, and from 480 to 660 feet above that of Ontario.

The Huron Tract was explored in 1827.[1]

[Dr. William "Tiger"] Dunlop, 1841.

NOTES

Preface

1. *Illustrated Historical Atlas of the County of Perth* (Toronto: H. Beldon and Co., 1879) ii.
2. William Johnston, *History of the County of Perth* (Stratford: The Beacon Office, 1903) 23.

A Note From the Author: Money in Canada 1763-1858

1. See D. McGillivray, "Money," in *The Canadian Encyclopedia* (3 volumes) (Edmonton: Hurtig Publishers, 1985) 1153-54. See also A.B. McCullough, *Money and Exchange in Canada to 1900* (Toronto: Dundurn Press and Parks Canada and the Canadian Publishing Centre, Supply and Services Canada Service, 1984) 292. See also Francis A. Evans, *Emigrants Directory and Guide to Obtain Lands and Effect a Settlement in the Canadas* (Dublin: William Curry, Jun. and Co.; London: Simpkin and Marshall; Edinburgh: Oliver and Boyd, 1833). Also from http://freepages.genealogy.rootsweb.com/~wjmartin/emigrant.htm and from Lutzen Redstra, Archivist-Administrator, Stratford-Perth Archives, Stratford, Ontario.

Chapter 1: Why The Canada Company?

1. Quote from the AO Robinson Papers: Comments made by John Beverley Robinson in March 1853 regarding a paper (with related minutes) delivered by him to British Colonial Undersecretary R.J. Wilmot on January 10, 1823; hereinafter referred to as "Robinson Paper – Comments." For a comprehensive overview of John Beverley Robinson's significant contribution in Upper Canada, see Robert E. Saunders, "Robinson, Sir John Beverley," in *Dictionary of Canadian Biography, Volume ix* (Toronto: University of Toronto Press, 1976) 668-679.
2. The Clergy Reserves in Upper Canada were a source of much friction. The debate centred on whether the lands had been set aside for the Church of England exclusively,

or should include the Church of Scotland, the Roman Catholic Church, the Methodists and/or any other Christian church. The issue was not resolved until 1854 when the lands were secularized.

3. Sir Peregrine Maitland was born in England in 1777, entered the British army at age 15 as an ensign and rose through the ranks to become a major general. He served with distinction in the Battle of Waterloo and was knighted in 1815. In 1818, he was appointed lieutenant governor of Upper Canada and while initially popular, he was decidedly conservative and "resisted American and democratic tendencies." Most of the day-to-day administration of the province was carried out by Major George Hillier, his secretary and confidant. Maitland tried to make the land department of Upper Canada self-supporting and to devise a method to make the Crown Reserves productive. While British Colonial Secretary Lord Bathurst favoured naturalizing aliens (immigrants from the United States) and granting them the civil rights and privileges of British subjects, Maitland did not. Maitland favoured the Church of England as the established church in Upper Canada, and opposed the Reformers who wanted the lieutenant governor's advisors chosen from the ranks of the elected Legislative Assembly. Given his authoritarian and hierarchical view of society, he wanted the appointed Legislative Council to be dominant. That and other issues lead to intense confrontation with Reformers. Maitland left office in November 1828. Many believed it was because of complaints against him. See Hartwell Bowsfield, "Maitland, Sir Peregrine," in *Dictionary of Canadian Biography, Volume viii* (Toronto: University of Toronto Press, 1985) 596-605.

4. Sir Robert John Wilmot-Horton was born in England in 1784 as Robert John Horton. Educated at Eton and Oxford (M.A.), he was elected to the House of Commons in 1818 where he represented his borough until 1830. Horton was appointed undersecretary of state for war and colonies in 1821, held that position until 1828 and was knighted in 1831. He took a particular interest in the North American colonies given their importance. In compliance with his father-in-law's will, Horton changed his surname to Wilmot-Horton on May 8, 1823. Although commonly referred to as Wilmot-Horton from then on, Canada Company directors merely called him "Horton" in their minute books and internal correspondence. See Philip Buckner, "The Colonial Office and British North America," in *Dictionary of Canadian Biography, Volume viii* (Toronto: University of Toronto Press, 1985) xxiii-xxxviii. See also *Dictionary of National Biography: From the Earliest Times to 1900* (London: Oxford University Press, published since 1917) 1284-85.

5. Lord Bathurst was secretary of state for the colonies from 1812 to 1827. He has been described as "reasonably intelligent and capable ... deeply conservative in his political philosophy but prepared to deal with specific issues pragmatically." He was viewed as being past his prime by the 1820s, which meant that Colonial Undersecretary Wilmot-Horton became quite powerful and, in fact, it is said that Wilmot-Horton overshadowed Bathurst on occasion. Lord Bathurst is alleged to have commented to a departing governor " Joy be with you, and let us hear as little of you as possible." See Philip Buckner, "The Colonial Office and British North America," in *Dictionary of*

Canadian Biography, Volume viii (Toronto: University of Toronto Press, 1985) xxiii-xxxvii.

6. See Chapter 2, "The Galt Era" for a comprehensive outline of John Galt's life before his involvement with the Canada Company.

7. AO, Robinson Papers, Paper delivered by J.B. Robinson to R.J. Wilmot, January 10, 1823; hereinafter referred to as "Robinson Paper."

8. Ibid.

9. Ibid.

10. R.R. Palmer and Joel Colton, *A History of the Modern World*, Third Edition (New York: Alfred A. Knopf, 1966) 410.

11. Aileen Dunham, *Political Unrest in Upper Canada 1815-1836* (Toronto: McClelland & Stewart, 1963) 47.

12. In October 1820, the price had fallen to fifty-four shillings. See "Upper Canada, Journals of the House of Assembly, 1818-1821," contained in the *Tenth Report of the Bureau of Archives for the Province of Ontario, 1913*, 431.

13. The suspension of the receipts from Lower Canada had resulted from a dispute relating to the division of customs duties collected in Lower Canadian ports. Since these duties were the chief source of revenue, they were of great importance to Upper Canada. Under arbitration proceedings in 1824, population was accepted as the basis of division and Upper Canada was allowed one-fourth of the net revenue until 1832 when the proportion was increased to one-third.

14. AO, Robinson Papers, Paper (with related minutes) delivered by J.B. Robinson to R.J. Wilmot, January 10, 1823.

15. Ibid.

16. Ibid.

17. See J.M. Cameron, "Guelph and the Canada Company 1827-1851: An Approach to Resource Development," unpublished M.Sc. thesis, University of Guelph, 1966. See also Lucille H. Campey's books on Scottish immigration/colonization schemes, all published in Toronto by Natural Heritage Books: *"A Very Fine Class of Immigrants": Prince Edward Island's Scottish Pioneers 1770-1850* (2001); *"Fast Sailing and Copper-Bottomed": Aberdeen Sailing Ships and the Emigrant Scots They Carried to Canada 1774-1855* (2002); *The Silver Chief: Lord Selkirk and the Scottish Pioneers of Belfast, Baldoon and Red River* (2003); and *After the Hector: The Scottish Pioneers of Nova Scotia and Cape Breton, 1773-1852* (2004).

18. Lloyd G. Reynolds, *The British Immigrant: His Social and Economic Development of Canada* (Toronto: Oxford University Press, 1936) 30-31. See also Donald MacKay, *Flight From Famine: The Coming of the Irish to Canada* (Toronto: McClelland & Stewart, 1990).

19. In 1818, township representatives of the Midland District, which includes the counties of Hastings, Prince Edward, Addington and Leeds, and Frontenac in today's terms, petitioned for the selling of government land in an orderly fashion. They suggested the revenue could be used to meet the claims of those who had suffered from the war, to improve the province more generally and to increase, eventually, Britain's revenue. See O.D. Skelton, *The Life and Times of Sir Alexander Tilloch Galt* (Toronto: Oxford University Press, 1920) 4.

20. Robinson suggested a lease-purchase plan which would enable a lessee to have his lease payments applied to the purchase price should he decide to buy the property being leased. It was not unlike the one implemented by Canada Company commissioner Frederick Widder some nineteen years later in 1842.

21. Robinson Paper – Attached Minutes.

22. Ibid.

23. For an account of Colonel Clark's life see Bruce A. Parker and Bruce G. Wilson, "Clark (Clarke), Thomas" in *Dictionary of Canadian Biography, Volume vi* (Toronto: University of Toronto Press, 1987) 147-50.

24. H.B. Timothy, *The Galts: A Canadian Odyssey: John Galt 1779-1839* (Toronto: McClelland & Stewart, 1977) 51.

25. For an account of William Dickson's life see Bruce G. Wilson, "Dickson, William," in *Dictionary of Canadian Biography, Volume vii* (Toronto: University of Toronto Press, 1988) 250-52. In 1827, Dickson renamed Shade's Mill (the first community founded on his lands by Absalom Shade) "Galt" in honour of his cousin. Today the community is part of Cambridge, Ontario.

26. Timothy, *The Galts: A Canadian Odyssey: John Galt 1779-1839*, 52.

27. Robinson Paper – Comments.

28. See G.M. Craig, "Strachan, John," in *Dictionary of Canadian Biography, Volume ix* (Toronto: University of Toronto Press, 1976) 751-766, for an account of the life of John Strachan – teacher, clergyman, office holder and bishop – who was larger than life. He was a fighter for what he believed in. He pushed successfully for legislation relating to common schools and, as a teacher trained many young men who went on to become leaders in Upper Canada in the second quarter of the 19th century. He believed in the connection between church and state and the status of the Church of England. Committed to the training of Church of England clergy, he established the Diocesan Theological Institution at Cobourg, Upper Canada, in 1838. Strachan was also a humanist. During the cholera epidemics of 1832 and 1834, for example, he was tireless in ministering to the sick, arranging for burial of the dead and trying to prevent panic.

   In 1827, he obtained a royal charter for the University of King's College at York (but it did not open until 1843). The college was distinctly Church of England in outlook; however, much to Strachan's chagrin, it came under government control in 1849 and was secularized the same year. In 1850, it became the University of Toronto. This led Strachan to secure a charter and funding for the establishment of another church-sponsored post-secondary institution, the University of Trinity College (now also part of the University of Toronto). It received its provincial charter in 1851 and opened its doors in 1852. The Diocesan Theological Institution was absorbed into it.

   Strachan was appointed bishop of the Toronto Diocese of the Church of England in 1839 at the age of 61. In 1842, he founded the Church Society, whose role was to watch over church finances, raise money for missionary activity and provide pensions for retired clergy. The Clergy Reserves issue was a preoccupation of his and he was quite disgusted when the Reserves were secularized in 1854. That said, he did manage to ensure that clergy stipends then being paid would still come from proceeds of the sale of the reserves out of a fund administered by the Church Society. Strachan only gave up his duties as bishop in 1866 at age 88.

Throughout his long life, Strachan sought to advance the economies of British North America for the betterment of society. To quote Craig, " Strachan has been correctly remembered as the arch tory of his era… He believed in an ordered society, an established church, the prerogative of the crown, and prescriptive rights; he did not believe that the voice of the people was the voice of God. No man could have fought harder or more persistently for what he understood to be right.…"

John and Ann Strachan had nine children: James McGill (born in 1808); Elizabeth (in 1810, died in 1814); George Cartwright (in 1812); Elizabeth Mary (in 1814 – she married Canada Company commissioner Thomas Mercer Jones in 1832); John (in 1815); Alexander Wood (in 1817); two daughters in 1821 and 1824 (they died as infants); and Agnes (in 1822 who died before reaching the age of 17). Strachan married well. Ann "had a great share of beauty" and an annuity for life of £300. For much of his life, he struggled financially because he liked to live well, wanted to provide properly for his children and donated to many charitable causes and good works. Ann Strachan died in 1865. The bishop died in 1867 and was given an impressive state funeral in Toronto.

29. LAC, Colonial Office Original Correspondence, Secretary of State, CO42/396, Galt to Bathurst, February 17, 1824. One suspects that Galt intended to be that special agent.

30. Ibid, Galt to Bathurst, March 8, 1824. Thomas Smith, sometime accountant of the Canada Company, also takes credit with Galt for the idea of forming the company. "The first idea of forming the Company was started in February, 1824 …. He [Galt] consulted me on the subject [of the reserves] and we agreed that it was a good and legitimate object for a public Company." See Thomas Smith, *An Address to the Shareholders of the Canada Company* (London: 1829) 4.

31. Ibid, Galt to Lord Bathurst, March 9, 1824.

32. Ibid, Horton to Galt, March 9, 1824.

33. For further background on Peter Robinson, see Wendy Cameron, "Robinson, Peter," in *Dictionary of Canadian Biography, Volume vii* (Toronto: University of Toronto Press, 1988) 752-57.

34. LAC, Colonial Office Original Correspondence, Secretary of State, CO42/396, Galt to Horton, March 2, 1824.

35. Robinson Paper.

36. G.A. Wilson, "The Political and Administrative History of the Upper Canada Clergy Reserves, 1790-1854," unpublished Ph.D. thesis, University of Toronto, 1959, 119.

37. For more on Alexander Macdonell, see J.E. Rea, "McDonell, Alexander," in *Dictionary of Canadian Biography, Volume vii* (Toronto: University of Toronto Press, 1988) 544-551. Born in Glengarry, Scotland, in 1762, Father Macdonell had organized the Glengarry Regiment in Scotland, of which he was chaplain. The regiment saw service in Guernsey and Ireland, but was disbanded in 1802 following the Peace of Amiens. Two years later, with the promise of 200 acres for every soldier from the regiment who emigrated, he arrived in Glengarry County in eastern Upper Canada with a

large group of his former soldiers to settle there. He became involved in the colonial election of 1812 and raised a regiment to fight in the War of 1812-14 (he was their chaplain). He also was very active in establishing parishes in the colony, eventually obtaining financial assistance from the colonial and British governments for priestly salaries and the building of churches.

Macdonell was therefore very familiar with conditions in Upper Canada when Galt sought him out. They were to become close friends, and Macdonell would not only receive some stock in the Canada Company for the helpful information provided, including a recent report on the Crown Lands Department, but would be given a prime site in Guelph by Galt for a Roman Catholic church as well. Macdonell went on to become a member of the Executive Council in 1831 – and as such was subject to the attacks by Reformers during that decade, which would draw him and Strachan together. In fact, Macdonell even wrote to his cousin, Colonial Secretary Lord Glenelg, in support of Strachan's becoming Anglican bishop of Toronto. Macdonell and Strachan differed on the Clergy Reserves issue in that Macdonell wanted his church to have a share in them.

Alexander Macdonell died in Scotland in 1840 as he was endeavouring to raise funds to build a seminary. The town of Alexandria in Glengarry County of eastern Ontario is named after the bishop.

38. LAC, C042/396, Galt to Horton, March 9, 1824.

39. Ibid, outline of proposed Canadian Land Company, April 28, 1824.

40. AO, Canada Company Papers, *Minutes of the Court of Directors*, Galt to Bathurst, July 8, 1824. Interestingly, in developing his concept for the Canada Company, Galt used much of the same reasoning regarding colonization as Lord Selkirk employed in implementing his much earlier settlement ventures in Prince Edward Island (Belfast, 1803), in Upper Canada (Baldoon, 1804, now Wallaceburg, Ontario) and in the later Red River settlement in Manitoba (1812). As an private individual of means, Selkirk spent about £114,000 (about £7 million in today's value) on his Scottish settlements in British North America. See Lucille H. Campey, *The Silver Chief: Lord Selkirk and the Scottish Pioneers of Belfast, Baldoon and Red River* (Toronto: Natural Heritage, 2003).

41. For more information on Thomas Talbot, see Alan G. Brunger, "Talbot, Thomas," in *Dictionary of Canadian Biography, Volume viii* (Toronto: University of Toronto Press, 1985) 857-62.

42. For more information on William Gilkison, see D.E. Fitzpatrick, "Gilkison, William," in *Dictionary of Canadian Biography, Volume vi* (Toronto: University of Toronto Press, 1987) 285-286.

43. But, in fact, the final terms were not the same. Two land companies received land grants in Australia. One, the Australian Agricultural Company in New South Wales (Upper Hunter Region), received a grant of one million acres in 1824, with the mandate to produce agricultural goods, including fine wool. The other, the Van Diemens Land Company in Tasmania, received its charter in 1825. Its mandate included the production of wool. It used indentured labour from the depressed rural

areas of England, Scotland and Ireland. Most of them were illiterate builders, shepherds, labourers and blacksmiths. The men would serve the company under contract for a set term in return for passage to the colony, food, housing and wages. In reality, food rations were inadequate, there was no housing on arrival and wages were low. See http://www.beswick.info/besfam/cottrell.htm and http//vision.net.au/~indenturedservants/oview.htm.

44. LAC, CO42/396, Horton to Galt, May 21, 1824.

45. AO, Canada Company Papers, *Minutes of the Court of Directors*, 7. Proceedings relative to the formation of the company are described in the Minutes of the General Meeting on July 30, 1824.

46. AO, Canada Company Papers, *Minutes of the Court of Directors*, July 5, 1824, Galt to Bathurst, June 28, 1824.

47. Ibid, Horton to Galt.

48. Ibid.

49. Ibid.

50. AO, Canada Company Papers, *Minutes of the Court of Directors*, Galt to Bathurst, July 8, 1824.

51. Ibid.

52. Ibid, Bathurst to Galt.

53. Ibid, September 17, 1824. From notes drawn up by Lieutenant Colonel Francis Cockburn who would be head of the five-man commission to ascertain a fair price for the land.

54. For more on Simon McGillivray, see Fernand Ouellet, "McGillivray, Simon," in *Dictionary of Canadian Biography, Volume vii* (Toronto: University of Toronto Press, 1988) 561-62. See also AO, Canada Company Papers, *Minutes of the Court of Directors*, July 6, 1827.

Simon McGillivray (c. 1783-1840) was a businessman and leading partner in the North West Company. In 1821, with Edward Ellice, he devised a solution to merge the XY Company/North West Company and the Hudson's Bay Company into the HBC. He was a member of the joint HBC/NWC board which was established to manage the fur trade according to the new configuration. The joint board was dissolved in 1824 just as the Canada Company's Provisional Committee was being established that year. McGillivray was a founding director of the Canada Company and a key player in the original and ongoing negotiations with the British government. At the invitation of his fellow directors, he became chairman of the company's Committee of Correspondence in April 1827, which was to be for a two-year term at an annual salary of £1,000. On accepting this appointment, he stated that he would only consider it "temporary" and to be held while he enjoyed the confidence of the directors.

In 1837 at age 52, McGillivray married Anne Easthope, the daughter of business partner and former Canada Company director, John Easthope. McGillivray was the provincial Grand Master of the Masonic movement in Upper Canada for 18 years and, in 1829, he helped organize the United Mexican Mining Association of London. In 1835, with John Easthope, he purchased the *Morning Chronicle and London Advertizer* newspaper.

55. Edward "Bear" Ellice, MP (1783-1863) was heir to a substantial fortune, inherited from his Scottish-born father, including very significant land holdings and commercial interests in the United States, British North America and the West Indies, was a highly successful businessman who was the power behind the amalgamation in 1821 of the XY Company/North West Company and the Hudson's Bay Company (HBC). He was deputy governor of the HBC, a member of the Canada Company's Provisional Committee, deputy governor of Canada Company and a member of company's Committee of Correspondence. Per article 15 of the original charter, as one-third of the directors were to be rotated off the board each year commencing in 1829, he left the Canada Company board in March 1829. As a politician in the 1820s, he was the principle spokesman for the Whigs on economic questions. In the early 1830s, he became secretary of the treasury and chief government whip. In 1833, he entered the cabinet as secretary of war. He left politics in 1834 to manage his substantial business interests.

    In the meantime, Ellice remained loyal to John Galt during Galt's difficulties following his return to England and release from debtor's prison. Together they promoted the idea of a land settlement company in the Eastern Townships of Quebec which became the British American Land Company headquartered in Sherbrooke, Quebec. Galt was its secretary for ten months in 1832, but ill health forced his resignation. The company received its charter in 1834 and had holdings in the Townships of 850,000 acres for resale to settlers.

    For more on Edward Ellice, see Note #74, Chapter 1 and Chapter 3, p. 85-87; and also James M. Colthart, "Ellice, Edward," in *Dictionary of Canadian Biography, Volume ix* (Toronto: University of Toronto Press, 1976) 233-39. See also AO, Canada Company Papers, *Minutes of the Court of Directors*, August 24, 1826.

56. LAC, CO42/396, McGillivray to Horton, August 30, 1824.

57. For more on Lieut. Col. Francis Cockburn, see Ed McKenna, "Cockburn, Sir Francis," in *Dictionary of Canadian Biography, Volume ix* (Toronto: University of Toronto Press, 1976) 132-34.

58. For more information, see Philip Buckner, "Harvey, Sir John," in *Dictionary of Canadian Biography, Volume viii* (Toronto: University of Toronto Press, 1985) 374-84.

59. As fifth commissioner, the company proposed one of the following: the Honourable John Richardson of Montreal, Samuel Gerrard of Montreal or John Davidson "of Quebec, now London." Davidson was chosen because of his being in London at the time. Galt was upset at being commissioner number four and wrote to Lord Bathurst, "I told Mr. Horton that I expected to be placed at the head of the Commission." LAC, Co42/396, Galt to Bathurst, December 3, 1824.

60. Per the minutes of the first Canada Company Court (Board) of Directors' meeting, which was held on July 30, 1824, it was unanimously resolved that the following would hold positions in the company, which would be re-confirmed following the granting of the Canada Company charter on August 19, 1826.

Directors:	Robert Biddulph, Esquire	Richard Blanshard, Esquire
	Robert Downie, Esquire	John Easthope, Esquire
	Edward Ellice, Esquire	John Fullarton, Esquire

Charles David Gordon, Esquire

William Hibbert, Esquire

John Hodgson, Esquire

John Hullett, Esquire

Hart Logan, Esquire

Simon McGillivray, Esquire

James McKillop, Esquire

John Mastermen, Esquire

Martin Tucker Smith, Esquire

Henry Usborne, Esquire

William Williams, Esquire

Chairman: Charles Bosanquet, Esquire

Deputy

Chairman: William Williams, Esquire

Auditors: Thomas Harling Benson, Esquire   Thomas Poynder, Esquire

Thomas Wilson, Esquire   John Woolley, Esquire

Secretary: John Galt, Esquire

Solicitors Messrs. Freshfield and Kaye

Bankers: Messrs. Masterman and Company, and Messrs. Cocks, Cocks, Ridge and Biddulph

At this meeting, it was also decided that four directors per year would rotate off of the Court of Directors commencing in 1829. See Appendix B: Canada Company Directors.

Most of the townships in the Huron Tract are named after the directors. See Appendix C: Huron Tract Township Names and Their Origins.

61. LAC, CO42/397, *Report of the Commissioners*, 2. The commissioners were paid in sterling as follows : Cockburn – £1,000; McGillivray – £1,000; Harvey – £650; Galt – £900; Davidson – £800.

62. J.B. Robinson warned Lord Bathurst that the company would be in a position to make a profit of from one to two hundred per cent "without having paid out a single farthing." Strachan thought the land was worth more. But no matter what happened, he wanted to ensure that the Clergy Corporation would control any sale and/or leasing of the lands. The Clergy Corporation also wanted to solve the issue as to whom the monies from the disposition of the lands should go before anything further was done. The Church of England claimed the Clergy Reserves while other denominations disagreed, with the Church of Scotland being the most vocal.

63. Robert Wilmot-Horton was perplexed and held as a private opinion that if, in fact, the company was in a position to make a profit as outlined by Robinson, one of two things must have happened. Either the commissioners' instructions were most unfortunately worded, or the commissioners misunderstood them in the process of arriving at a valuation.

64. AO, Canada Company Papers, *Minutes of the Court of Directors*, February 2, 1826, Horton to Bosanquet, January 30, 1826.

65. LAC, 0042/398, 384-406, Horton to the (five) commissioners, December 2, 1825, 4. (The page numbers referred to designates the *letter* page number.)

66. Ibid, 15.

67. Ibid, 18.

68. Ibid.

69. AO, Canada Company Papers, *Canada Company Correspondence*, Masterman to Bathurst, August 30, 1825.

70. LAC, CO42/398, Horton to Galt, July 26, 1825.

71. LAC, C042/398, Galt to Horton, December 18, 1825.

72. AO, Canada Company Papers, *Minutes of the Court of Directors*, May 5, 1826, Horton to Galt, May 5, 1826.

73. A search of various records has not revealed background on Sir Giffen Wilson, but it is thought that he was a lawyer. In any case, he must have been well-connected in government circles.

74. John Richardson was obviously chosen to represent the Canada Company because of his connection to the Ellice family and, more particularly, to Edward "Bear" Ellice who was a member of the Provisional Committee of the company. The Ellice family had had significant business interests in the United States since before the American Revolution. These included the fur trading partnership of the London, England–based Phyn, Ellice and Company (whose U.S. operation was headquartered in Schenectady, N.Y.) and significant land holdings in the United States and Canada which, at its peak, totalled 450,000 acres. The firm belonged to Edward's father, Alexander, and Edward's uncle, Robert.

    Richardson, a man of prodigious energy, had been in the employ of Phyn, Ellice and Company earlier. In 1783, at the age of 33, he rejoined the firm and was subsequently sent to Montreal in 1787 to help reorganize it. On Robert Ellice's death in 1790, Richardson became a partner and the firm was renamed Forsyth, Richardson and Company. The business of the company included fur trading, freight forwarding, ownership of vessels on the Great Lakes and the Atlantic and extensive landholdings in North America. The company also became a minor shareholder in the North West Company, was a co-partner in the XY Company and lobbied for political changes in the interests of commerce. Richardson was very involved politically in Lower Canada. He was elected to the Legislative Assembly of Lower Canada a number of times beginning in 1792. In 1811, he became a member of the province's Executive Council and held that seat until he died in 1831. After the War of 1812-14, his firm became further involved in exports, including grain and timber, and he pushed for the construction of the Lachine Canal. By this time, he was Edward Ellice's agent to oversee the management Ellice's huge seigneury, "Beauharnois," on the south shore of the St. Lawrence River west of Montreal. As a member of the Canada Company's Provisional Committee, Ellice obviously thought Richardson to be an excellent choice to look out for the company's interests at this juncture. It is interesting to note that among Richardson's many other pursuits, he was an active promoter of the union of Upper and Lower Canada. For more information, see F. Murray Greenwood, "Richardson, John," in *Dictionary of Canadian Biography, Volume vi* (University of Toronto Press, 1987) 639-46.

75. It is believed that Canada Company board member and shareholder Charles Bosanquet is the same Charles Bosanquet as found in the British *The Dictionary of National Biography: From Earliest Times to 1900*, who was born in England in 1779 and died in

1850. Bosanquet successfully engaged in mercantile pursuits as a member of a Huguenot family of entrepreneurial London merchants and was a sub-governor, then governor, of the South Sea Company. From 1823 to 1836 he was chairman of the British Exchequer Bill Office, and was appointed lieutenant colonel, later colonel, of the Light Horse Volunteers. He authored a number of treatises on trade and monetary issues. *Dictionary of National Biography: From the Earliest Times to 1900* (London: Oxford University Press, published since 1917) 874.

76. LAC, Canada Company Papers 1825-1887, Horton to Bosanquet, February 9, 1826.

77. AO, Canada Company Papers, *Minutes of the Court of Directors*, May 8, 1826, Horton to Galt, May 5, 1826.

78. Ibid.

79. Thelma Coleman, *The Canada Company: with supplement by James Anderson* (Stratford: County of Perth and Cummings Publishers, 1978) 174.

80. Ibid.

81. Ibid. The story of the relationship between Aboriginals/First Nations/Native Canadians and European immigrants has been one of conflict, confusion and controversy. While these complex issues are beyond the scope of this book, the following references provide some context. See Edward S. Rogers and Donald A. Smith, eds., *Aboriginal Ontario: Historical Perspectives on the First Nation.* Ontario Historical Studies Series. (Toronto: Dundurn Press, 1994). (See Part II, Chapter 6, "Land Secessions, 1763-1830" by Robert J. Surtees, 92-121.) See also Arthur J. Ray, *I Have Lived Here Since The World Began: An Illustrated History of Canada's Native Peoples* (Toronto: Lester Publishing, Key Porter Books, 1996) and Robert MacDougall (Elizabeth Thompson, ed.), *The Emigrant's Guide to North America* (Toronto: Natural Heritage, 1998).

From all accounts, it would appear that the relationship (good or bad) that existed between the Native Peoples and the newcomer settlers depended on the "chemistry" between individuals and/or groups. A positive example would be the account handed down from Charles George Middleton through his son, Charles George Jr., to his son, William Harvey, the author's great-uncle. Charles George (21) married Elizabeth Wise (c. 20) in Kent in 1833, and, the next year, they both emigrated to York in Upper Canada. After his wife gave birth there to Charles Jr. in 1835, Charles Sr. walked alone to the Huron Tract and initially purchased 80 acres for £30 from the Canada Company. The land lay between Clinton and Bayfield in Goderich Township. (An additional 80 acres would be purchased in 1841 for £50.) He then returned to Toronto and brought his young family back to his homestead. They befriended the local Chippewa. One day, the Natives came looking for food. The Middletons were generous and shared what they had. They were well repaid the day Charles Sr. and Charles Jr. were driving their wagon over the corduroy road to get supplies and were stopped by a fallen tree, the result of a windstorm the previous night. The bush was dense and there was no way around the obstacle. Seemingly out of nowhere, some Chippewa appeared, removed the tree and were gone. The Middletons were able to continue on their way. Taken from Margaret Eleanor (Smith) Wilson, "A Summary of the History of Parts of the Middleton Family," not dated.

W.W. Smith's *Gazetteer* reported on a Chippewa settlement at the mouth of the Saugeen River north of the Huron Tract in the mid-1840s. It indicated that the "Indians" there appear to have "settled and converted" [to Christianity] about 1831 on 200-300 acres of good land – presumably having migrated north after the government purchase of their more southerly lands that had been acquired by the Canada Company. The resident missionary living among them reported that they cut sufficient hay for their oxen and cows, grow excellent corn and have "several good log houses and several comfortable bark shanties." There are also several fields of fine corn and potatoes. "The Indians in this station have been remarkable for their steadfastness since they embraced Christianity. They appear to be a happy people: much attached to their missionaries, teachable, and give solid proofs that they are progressing in civilization." On the other hand, the government's chief superintendent reported "they appeared very poor and very miserable, trusting very much to hunting and fishing for their support. The fishing is very productive, and has attracted the notice of the white people, who annoy the Indians by encroaching n [sic] what they consider their exclusive right, and on which they rely for orovisions [sic]. They hunt in the tract belonging to the Canada Company..." In 1836, the Chippewa had surrendered 1.6 million acres of land north and east of there, and got in return what is now the Bruce Peninsula which is "supposed to contain about 450,000 acres." See Wm. H. Smith, *Smith's Gazetteer* (Toronto: H. & W. Rowsell, 1846; Facsimile edition: Toronto: Coles, the Book People: 1970) 165-66.

82. For more information on John Rolph, see G.M. Craig, "Rolph, John," in *Dictionary of Canadian Biography, Volume ix* (Toronto: University of Toronto Press, 1976) 683-90.

83. AO, Canada Company Papers, *Minutes of the Court of Directors*, May 20, 1826, Rolph to Galt, May 13, 1826.

84. Ibid.

85. AO, Canada Company Papers, *Minutes of the Court of Directors*, May 30, 1826, Arrangement Made and Concluded between Lord Bathurst and the Canada Company, May 23, 1826.

86. AO, Canada Company Papers, *Minutes of Committees*, Committee to Prepare a Report for the General Meeting, July 7, 1826.

87. Ibid, AO, Canada Company Papers, *Minutes of the Court of Directors*, May 30, 1826.

88. AO, Canada Company Papers, *Minutes of Committees*, Committee to Prepare a Report for the General Meeting, July 7, 1826.

89. Ibid.

90. AO, Canada Company Papers, *Minutes of the Court of Directors*, June 16, 1826 and July 18, 1826.

Chapter 2: The Galt Era

1. Robina and Kathleen Macfarlane Lizars, *In the Days of the Canada Company 1825-1850* (Toronto: William Briggs, 1896), Introduction by G.M. Grant, vi.

2. Information on John Galt drawn from Roger Hall and Nick Whistler, "Galt, John," in *Dictionary of Canadian Biography, Volume vii* (Toronto: University of Toronto Press,

1988) 335-40. See also H.B. Timothy, *The Galts: A Canadian Odyssey: John Galt 1779-1839* (Toronto: McClelland & Stewart, 1977) 45-6.

3. The Canada Company used the terms "superintendent" and "commissioner" interchangeably in correspondence.

4. AO, Canada Company Papers, *Minutes of Committees*, Committee of Correspondence, September 25, 1826.

5. AO, Canada Company Papers, *Minutes of the Court of Directors*, September 4, 1826.

6. Ibid, October 3, 1826.

7. AO, Canada Company Papers, *Minutes of Committees*, Committee of Correspondence, September 18, 1826.

8. Ibid, August 26, 1826.

9. LAC, Canada Company Papers, 1825-1887, *Canada Company Prospectus*, 1824.

10. AO, Canada Company Papers, *Minutes of Committees*, Committee of Correspondence, April 19, 1827.

11. Ibid.

12. Ibid.

13. For further detail on the fascinating William "Tiger" Dunlop, see Gary Draper and Roger Hall, "Dunlop, William," in *Dictionary of Canadian Biography, Volume vii* (Toronto: University of Toronto Press, 1988) 260-64.

14. AO, Canada Company Papers, Minutes of the Court of Directors, September 11, 1826, "First Report of the Committee of Correspondence – Instructions to the Warden of the Forests."

15. John Galt, *The Autobiography of John Galt, Volume ii* (London: Cochrane and M'Crone, 1833) 4, 5.

16. The "Minutes of the Court of Directors" record the following report of the company's General Committee (whose role was "to consider of all matters and things which may be expedient for the Company to adopt"): "Mr. William Dunlop has been mentioned as a Gentleman singularly well qualified for the tasks of activity and enterprise here proposed. It has been stated that he is well acquainted with the localities of Upper Canada and particularly with those of the wilder and remoter parts of the located districts having been employed in the opening of the Military roads and particularly the Penetanguishene Road between Lake Simcoe and Lake Huron and that he was also employed as the Manager in the laborious undertaking of clearing the jungles of the Island of Sauger at the entrance of the Ganges. Thus it has happened that by a fortuitous combination of circumstances your Committee are enabled to recommend for a class of multifarious and laborious duties a Gentleman possessed of peculiar qualifications both in experience and habits." AO, Canada Company Papers, Minutes of the Court of Directors, September 11, 1826.

17. Ibid.

18. Frederick H. Armstrong and Ronald J. Stagg, "Mackenzie, William Lyon," in *Dictionary of Canadian Biography, Volume ix* (Toronto: University of Toronto Press, 1991) 496-509.

19. These young men were the sons of prominent, well-connected families who did not appreciate Mackenzie's attacks on the Tory establishment and what they represented.

20. Hartwell Bowsfield, "Maitland, Sir Peregrine," in the *Dictionary of Canadian Biography,*

Volume viii (Toronto: University of Toronto Press, 1985) 596–605.

21. John Galt, *The Autobiography of John Galt, Volume ii*, 66.

22. Ibid, 66, 67.

23. The Court of Directors met regularly to deal with routine and other issues relating to the operation of the company. Not all directors attended all meetings. Generally this did not appear to cause great concern, however it was a somewhat troublesome issue to ensure that all directors were kept informed of company matters. The Minutes of the July 6, 1827, meeting reveals that two months earlier, a discussion about "Attendance Money" for directors had taken place at which time it was proposed that attendees should each receive £18 per meeting. This was rejected, but a £15 payment was approved. To simplify matters, the company accountant would take the requisite funds to the actual meeting and pay the directors on the spot.

   A quorum was five. Typically five to seven directors would attend meetings. AO, Canada Company Papers, *Minutes of the Court of Directors,* July 6, 1827.

24. AO, Canada Company Papers, *Minutes of the Court of Directors*, November 5, 1827.

25. LAC, Miscellaneous Records Relating to the Canada Company, *Minutes of the Executive Council,* Vol. 41, March 5, 1828, Hullet to Galt, November 5, 1827.

26. AO, Canada Company Papers, *Minutes of the Court of Directors*, November 5, 1827.

27. AO, Canada Company Papers, *Correspondence with His Majesty's Government*, McGillivray to Huskisson, September 13, 1827.

28. Ibid, Stanley to McGillivray, October 25, 1827.

29. AO, Canada Company Papers, *Correspondence with His Majesty's Government*, Maitland to Bathurst, October 17, 1827.

30. Whether the settlers had paid ten shillings per Maitland's comment, or eight shillings per the settlers' complaint, the company stood to make a nice profit either way. Miscellaneous Records Relating to the Canada Company, *Minutes of the Executive Council, Vol. xli,* December 29, 1827, "The Petition of the Scotch Settlers in Guelph late from La Guayra," dated December 12, 1827. The petition resulted in an exhaustive inquiry, but in the end, the settlers did not obtain a grant of land.

31. Ibid, Galt to Hillier, December 26, 1827.

32. Ibid, Hillier to Galt, December 27, 1827.

33. Ibid, December 31, 1827.

34. The British *Fraser's Magazine* of November 1830 carried a description of the founding of Guelph "this second Rome or Babylon" based on a letter Galt had written to a friend on June 2, 1827: "The site chosen was on 'a nameless stream's untrodden banks,' about eighteen miles in the forest from GALT .... Early on in the morning of St. George's Day, I proceeded on foot towards the spot, having sent forward a band of woodsmen with axes on their shoulders to prepare a shanty for the night – a shed made of boughs and bark, with a great fire at the door. I was accompanied by my friend Dunlop, a large fat, facetious fellow of infinite jest and eccentricity, but he forgot his compass, and we lost our way in the forest. After wandering up and down like babes in the woods – the rain raining in jubilee – ... we came to a hut of a Dutch settler ... We hired him for our guide.

   It was almost sunset when we arrived at the rendezvous; my companion, being wet to the skin, unclothed and dressed himself in two blankets, one in the Celtic and the other in the Roman fashion – the kilt and the toga ... "I kept my state" (as Macbeth says of his wife at the banquet) of dripping drapery. We then with surveyors and woodmen ... proceeded to a superb maple tree, and I had the honour and glory of laying the axe to the root thereof, and it soon fell "beneath our sturdy strokes" with the noise of an avalanche. It was the genius of the forest unfurling its wings and departing forever. Being the king's name-day, I called the town Guelph – the smaller fry of the office having monopolized every other I could think of; and my friend drawing a bottle of whiskey from his bosom, we drank prosperity to the unbuilt metropolis of the new world." Robina and Kathleen Macfarlane Lizars, *In the Days of the Canada Company 1825-1850* (Toronto: William Briggs, 1896) Appendix, 481-82.

35. AO, Canada Company Papers, *Minutes of the Court of Directors*, July 6, 1827.

36. See Jim Cameron and Mike Cowles, *The Goderich Signal-Star,* "Celebrating 175 years," June 26, 2002, 1-4, for an article on Lord Goderich and the founding of the town named after him: "We came to a beautiful anchorage of water in an uncommonly pleasant small basin selected by the Doctor."

The account of Frederick John Robinson in the British *Dictionary of National Biography: From Earliest Times to 1900* gives a very fulsome overview of his life. Born in London in 1782, Robinson was the second son of Baron Grantham. He was educated at Harrow and Cambridge, ran for election in 1806 and sat in the House of Commons as a moderate Tory for 20 years. Robinson held a wide variety of important positions: a lord of the admiralty (1810); president of the board of trade (1818-23; 1841-43); chancellor of the exchequer (1823-27); secretary of state for war and the colonies (1827; 1830-33); leader of the House of Lords (1827); prime minister (August, 1827 to January, 1828); lord privy seal (1833-34) and president of the India Board (1843-46). He was also the first president of the Royal Geographical Society (1830-33), a trustee of the National Gallery, president of the Royal Society of Literature and an elected fellow of the Royal Society. Robinson was created Viscount Goderich of Nocton on April 28, 1827 and Earl of Ripon on April 13, 1833.

The *Dictionary of National Biography: From Earliest Times to 1900* comments: "Ripon was an amiable, upright, irresolute man of respectable abilities and business like habits. The sanguine view in which he indulged while chancellor of the exchequer ... led him to be nicknamed 'Prosperity Robinson' while for his want of vigour as secretary of the colonies he received ... the name 'Bloody Goderich.' Though a diffuse speaker and shallow reasoner, he was said to have had the requisite skill to enliven dry subjects of finance 'with classical allusions and pleasant humour' which made them acceptable to most. In the House of Commons, he attained a certain popularity, but on his accession to the House of Lords, his courage and his powers alike deserted him. His want of firmness and decision of character rendered him to be quite unfit to be leader of a party in either house." See *Dictionary of National Biography From Earliest Times to 1900, Volume xvii* (London: printed since 1917) 7-11.

W.D. Jones, the biographer of Lord Goderich, is more charitable about the former prime minister and notes his "careful regard for the views and feelings of other people, his continuous efforts to apply the Golden Rule in his relations with his fellow men, his studied effort to purge all resentments and ill-feelings from his mind, and his anxiety at all times to avoid conflicts, set him apart from the overwhelming majority of other men who have made their mark in Western politics." See William Devereux Jones, *Prosperity Robinson: The Life of Viscount Goderich 1782-1859* (London: Macmillan, 1967) 279.

Lord Goderich was a friend of the Canada Company whom the directors wanted to recognize, primarily because of his support in the House of Commons for the Canada Company legislation which lead to the granting of the company's charter in 1826. His name lives on in the Town of Goderich and in Goderich Township.

37.   Following his run-in with the directors over the naming of Guelph, Galt wasted no time in founding Goderich. It was to be on the shores of Lake Huron at the mouth of the river which subsequently would be renamed the "Maitland" in honour of the lieutenant governor, Sir Peregrine Maitland. The river had been called the "Menesetung" by the Chippewa (meaning "laughing, healing, bubbling waters" or "berries heard before seeing" – rapids heard before being seen in the next winding of the river). On the other hand, early arrivals at the river mouth called it the "Red" because of the reddish colour of the water. These individuals would have included Royal Naval Captain Henry Wolsey Bayfield who completed an extensive marine survey of Lake Huron in the early 1820s and two traders from Detroit, W.F. Gooding and Frank Deschamps who set up camp there in the early 1820s to trade with the Chippewa.

Following the founding of Guelph, Tiger Dunlop and a survey crew made their way through the dense forests of the Huron Tract to Lake Huron. Once there, Dunlop built a log cabin overlooking the lake and the river mouth. He called it "The Castle." Galt now wanted to inspect the site and, to reach it, travelled from York via Lake Simcoe to Penetanguishene on Georgian Bay. He boarded the HM gunboat *Bee* which sailed around what is now referred to as the Bruce Peninsula and made his way south. On arrival at the mouth of the Menesetung River on June 27, he found Dunlop, his Castle and his surveyors/woodsmen. Two days later, on June 29, 1827, the two men would officially christen the townsite as "Goderich." The Canada Company directors had accepted the *fait accompli* and the naming issue was dropped.

Robina and Kathleen Lizars in the *In the Days of the Canada Company* recount the following: "He [Dunlop] selected, as the site for the new Canada Company town, a spot where Champlain was said to have halted while paddling and tramping his famous western journey of adventure. Here the Trader Gooding and one Frank Deschamp[s], a Frenchman, had already established a trading post and had built a hut close by the water's edge. Before this the Jesuit missionaries, whose beat was from ocean to ocean, had been the only white settlers. Sproat, MacDonald, McGregor from Zorra [Township] with his yoke of oxen, and others, now joined Dunlop; supplies arrived for them by water, and on top of the cliff facing the lake and bordering the river they built a small log house. This was known at once as 'The Castle' and Galt, in York, advised of its completion, made haste to visit his friend and give his sanction to the choice of situation. Already the workmen were there, ready to begin at word from him. His Majesty's gunboat, the *Bee*, was placed at Galt's disposal. ... Stories ... tell us of a romantic if not comfortable journey until, through a telescope, they made out a small clearing in the forest, and set on the brow of the cliff behind Dunlop's new-made castle. In a canoe set out to meet him, Galt found a strange combination of Indians, velveteens and whiskers, and discovered within the roots of the red hair the living features of the Doctor. ...The evening closed with a feast, and at it appeared ... champagne, presumably the first drunk at that very remote spot. A day was spent exploring the windings of the beautiful river."

See Robina and Kathleen Macfarlane Lizars, *In the Days of the Canada Company* (Toronto: William Briggs, 1896) 68-69.

38. AO, Canada Company Papers, *Minutes of the Court of Directors*, August 3, 1827. With the founding of Guelph and Goderich, Deputy Provincial Surveyor, John McDonald, would be responsible for the layout of the unique town sites of Guelph and Goderich in 1829 and later, in 1832, for Stratford. He also would be responsible for the surveys of all the townships in the Huron Tract (except Goderich Township) and, in 1828, was in charge of the survey of the Huron Road through dense woods from Wilmot Township to Goderich (now highway #8).

McDonald was born in May 1794, in Balachladdick, Parish of Dores, Inverness-shire, Scotland. Educated as an engineer in Scotland, he had emigrated with his family to the United States while still a young man. The family then moved to Upper Canada in 1823. Shortly after his arrival, he qualified as Deputy Provincial Surveyor, and then became associated with John Galt and Tiger Dunlop.

John McDonald would marry three times (his first two wives died) – Elizabeth Amelia Mitchell from Woodstock, Upper Canada, in 1832, and Mary Jane Carroll from Simcoe, Upper Canada, in 1850. Five of his six children with Elizabeth Amelia died in infancy, but the six born by Mary Jane survived to adulthood. His third wife, whom he married in 1861, was Mary Fraser of Inverness, Scotland. In 1845, McDonald was appointed sheriff of the District of Huron, following the death of the first sheriff, Henry Hyndman. McDonald served in that position until his death in 1873 at age 80.

39. Hilary Bates Neary, "Stanton, Robert," in *Dictionary of Canadian Biography*, *Volume ix* (Toronto: University of Toronto Press, 1976) 740-41.

40. Robert Lochiel Fraser, "Macaulay, John," in *Dictionary of Canadian Biography*, *Volume viii* (Toronto: University of Toronto Press, 1985) 513-22. The spelling of "Macauley" is not consistent in company and archival records. It varies from "Macauley" to "MacAuley" to "McAuley." The version used in this book is as found in the *Dictionary of Canadian Biography*, except where in quotation marks.

41. AO, "MacAulay" Papers, Robert Stanton to John "MacAulay," August 27, 1827.

42. AO, Canada Company Papers, *Minutes of Committees*, Committee of Correspondence, February 14, 1828.

43. See John Galt, *The Autobiography of John Galt*, II, 105, where he contradicts this statement. "Colonel Coffin, the head of the militia department, mentioned that his Excellency thought of appointing me to the command of a regiment."

44. Galt's actual salary was £1,000 sterling, plus an allowance for residence and expenses of £500 currency. In addition, a commission of 5% was to be paid on the actual cash received from sales and interest to a maximum of £500. Upon reaching that sum, the

allowance was to be discontinued, with future remuneration set at £1,000 sterling, plus commission of 5%. AO, Canada Company Papers *Minutes of Committees*, Committee of Correspondence, July 9, 1827.

45. AO, John Galt Papers, John Galt to Samuel Omay, November 20, 1827.

46. C.W. Robinson, *Life of Sir John Beverley Robinson* (Toronto: Morang and Co. Ltd., l904) 424.

47. AO, John Galt Papers, John Galt to Samuel Omay, November 20, 1827.

48. For an account of the life of John Walpole Willis and Lady Mary, see Alan Wilson, "Willis, John Walpole," in *Dictionary of Canadian Biography*, *Volume x* (Toronto: University of Toronto Press, 1972) 704-07.

49. Robert L. Fraser, "Baldwin, William Warren," in *Dictionary of Canadian Biography*, *Volume vii* (Toronto: University of Toronto Press, 1988) 35-44.

50. For an account of H.C. Thomson's life, see H.P. Gundy, "Thomson, Hugh Christopher," in *Dictionary of Canadian Biography*, *Volume vi* (Toronto: University of Toronto Press, 1987) 772-74.

51. For a fascinating account of Barnabas Bidwell's life, see G.H. Patterson, "Bidwell, Barnabas," in *Dictionary of Canadian Biography, Volume vi* (Toronto: University of Toronto Press, 1987) 54-58. Bidwell was born in Massachusetts in 1763. Trained as a lawyer, the former attorney general of that state and member of Congress fled to Upper Canada (Bath) in disgrace in 1810 because of certain less-than-honest financial dealings not of his making, but for which he was held responsible. The charge was much later refuted. He was a skilled "party organizer and manipulator of public opinion" and was elected in 1821 to the House of Assembly for Lennox and Addington, but was thrown out for what he called "ruthless tory opponents bent upon depriving all other unnaturalized persons of their civil rights." He did not run for office again but supported his son, Marshall, who became a most able member of the Assembly and a leader in the Reform Party.

52. For more detail on Marshall Bidwell, see G.M Craig, "Bidwell, Marshall Spring," in *Dictionary of Canadian Biography, Volume x* (Toronto: University of Toronto Press, 1972) 60-64. Marshall Spring Bidwell, the son of Barnabas Bidwell, was a successful court-room lawyer who represented Lennox and Addington in the Assembly from 1824 to 1836, after several attempts to have him disqualified because of his alien status as a settler from the United States. He supported legislation to improve the lot of electors which were regularly thrown out by the Legislative Council and, in 1828, had a leading role in the "alien" controversy which led to passage of an acceptable natural-ization act that year. He clashed with John Strachan on ecclesiastical issues. Bidwell went on to be speaker in the Assembly, but lost his election bid in 1836 and focused on the practice of law. Unwittingly, he got caught up in controversy around the events of 1837 and was "banished" to the U.S. in December of that year. He went on to become a successful lawyer in New York City.

53. The late Queen Mother (formerly Lady Elizabeth Bowes-Lyon) (1900-2002) was a descendant of Lady Mary Isabella's father, the 11th Earl of Strathmore. The Queen Mother's father was the 14th Earl of Strathmore.

54. Aileen Dunham, *Political Unrest in Upper Canada 1815-1836* (Toronto: McClelland &

Stewart, 1963) 114. In June 1828, as events transpired, Willis was removed from the bench for exercising poor judgment and returned to England, despite Galt and others encouraging him to stay in Upper Canada and seek redress in London. His wife, in the meantime, ran off with an infantry officer and they eloped to England, leaving her son in a maid's care. See Alan Wilson, "Willis, John Walpole," in *Dictionary of Canadian Biography, Volume x* (Toronto: University of Toronto Press, 1991) 704-06.

55. AO, Canada Company Papers, *Minutes of Committees*, Committee of Correspondence, Bosanquet to McGillivray, February 14, 1828.

56. See account in H.B. Timothy, *The Galts: A Canadian Odyssey: John Galt 1779-1839* (Toronto: McClelland & Stewart, 1977) 90. "Since the shanty put up by the woodmen on the campsite near the spot where the maple tree was cut down would not hold everyone, Galt ordered Prior [Charles Prior was the Canada Company's manager in Guelph] to have a lean-to of brushwood prepared to give shelter to some of the woodmen ... One by one the men left their places at the campfire and retired to their sleeping quarters. When Galt followed suit he discovered there was no room left in the shanty, so Prior invited him into the lean-to with, 'Welcome to the Priory.'" The Priory had become the name of Guelph's second building. It had various uses over the course of time including residence for John Galt and his family, a store, hotel, public chamber, receiving place for settlers and school, with part of it even used as a prison. In 1887, it was moved and became a railway station for the Guelph Junction Railway. In 1911, it was shifted to a vacant lot, and the structure deteriorated. In 1926, there was a small fire in the building which resulted in it being torn down. The original site is on what is now the Guelph Junction Railway about 48 metres (150 feet) west of the Macdonell Street intersection. See Thelma Coleman, *The Canada Company: with supplement by James Anderson* (Stratford: County of Perth and Cummings Publishers, 1978) 279-83.

57. C.A. Burrows, *The Annals of the Town of Guelph 1827-1877* (Guelph: 1877) 13.

58. AO, Canada Company Papers, *Minutes of the Court of Directors*, July 6, 1827. During this period, the Church of England in Canada tried a number of ways to get the company to contribute to the church and its causes. For example, company minutes report that the Bishop of Quebec tried to donate £100 to the company on condition that churches be built at company expense. The directors declined the gift, noting "receiving such contribution would imply some responsibility on the part of the Company to carry the object into effect." Archdeacon Strachan, for his part, tried to get the company to contribute to the university he was promoting at York. He was not successful. See AO, Canada Company Papers, *Minutes of the Court of Directors*, February 20, 1827, and August 3, 1827.

59. AO, Canada Company Papers, *Minutes of the Court of Directors*, March 20, 1828.

60. The April 15, 1828, *Minutes of the Court of Directors* are instructive as to how the Committee of Correspondence under Simon McGillivray's chairmanship viewed Galt's role in Canada. The Minutes also indicate that the directors felt they should tighten up procedures before sending Smith to Canada given the monies that Galt and Smith would be handling: "That in consideration of the amount of the company's funds which must be received by and placed at the disposal of the Functionaries

employed in Upper Canada, particularly the Superintendent and the Cashier, it is fit in all future appointments to require surety for the safe custody and the due application of such Funds. Considering however the manner in which the appointment of Superintendent has taken place. [Mr. Galt was originally sent as the Secretary on a mission of inquiry, which was then extended to experiment, and subsequently into the present system of administering the Company's transactions in that Country.] The Committee do not mean that security should *now* be required of *him*, but as the commencement of what appears to be a proper system for the future, they recommend ... That security ... to the extent of one thousand pounds be required from Mr. Thomas Smith on giving him the appointment of Cashier and Accountant."

For his part, Smith believed he would be able "to procure" £100 in bonds from each of "five respectable Friends." In the end, four friends posted bonds of £100 each, while three other persons connected to the company each posted £200 bonds, including directors Simon McGillivray and John Hullett, and accountant Robert Auld. AO, Canada Company Papers, *Minutes of the Court of Directors*, April 28, 1828.

61. What Galt did not know was that Smith had pushed for the hiring of one or more clerks in Canada with one of them reporting directly to him as an "Assistant in the Department of Accounts." The Committee of Correspondence agreed to this suggestion, but the Court of Directors did not, signifying that it would be their prerogative to so decide. It is interesting to note here that there was considerable discussion amongst the directors about Smith's role. AO, Canada Company Papers, *Minutes of the Court of Directors*, May 1, 1828.

62. AO, Canada Company Papers, *Minutes of Committees*, Committee of Correspondence, February 21, 1828, memorandum in regard to the state of the Company's Capital and Finances.

63. AO, Canada Company Papers, *Correspondence with His Majesty's Government*, McGillivray to Hay, February 23, 1828.

64. "Lumberers" were lumberjacks who went into the forest during the winter and cut down timber of any value for export. They could have been working independently or on behalf of known timber exporting companies. The reserves had suffered from illegal logging more than the other lands in this regard because there was doubt as to the right of possession.

65. AO, Canada Company Papers, *Minutes of the Court of Directors*, April 3, 1828, Agreement with His Majesty's Government, May 28, 1828. The Minutes of the Court of Directors of April 3, 1828, spell out how the issue of land unfit for cultivation, including swamps, ponds and lakes in the Huron Tract, was to be dealt with. In the second agreement between the British government and the company pertaining to lands to be purchased by the company, it was decided that an arbitrator would be appointed to deal with this issue. Subsequently, it was agreed by both parties that the size of the Tract would be increased from one million acres to "eleven hundred thousand acres, with the distinct understanding that the North Eastern boundary line of the tract as at present drawn includes not less than 50,000 acres of Swamp, or Lakes or Ponds situated in such Swamp or Land as unsaleable and wholly valueless to ordinary

settlers..." It was further agreed that if that swamp and "unsaleable" land within the North Eastern boundary are found to be less than 50,000 acres, the company would pay the difference in the same manner as if the Huron Tract was found to contain more than 1,100,000 acres, the company would pay for the excess land at a price fixed for the other lands in the Tract. If the land area were to fall short, "a rateable reduction" was to be allowed the company. That ended the discussion about "overages and shortages." AO, Canada Company Papers, *Minutes of the Court of Directors,* April 3, 1828.

66. AO, Canada Company Papers, *Correspondence with His Majesty's Government,* McGillivray to Huskisson, October 26, 1827.

67. Ibid, December 27, 1827.

68. AO, Canada Company Papers, *Minutes of the Court of Directors*, May 1, 1828.

69. Ibid.

70. The Lizars have an account of Messrs. Galt and Dunlop's "sense of humour" and how they "got back at" Smith which is indicative of the depth of ill feeling which existed between the accountant and the Dunlop/Galt twosome. Smith, in the words of the authors, was "the bad tempered Accountant mentioned very bitterly in Galt's 'Autobiography,' 'devoured by vanity,' 'full of airs and arrogance,' 'by this time an affliction'; and doubtless Dunlop felt he was now paying off some of his friend's grievances." This included causing the mare which Smith rode one day to accompany Galt and Dunlop, to bolt, requiring the accountant to hold on for dear life; much to Galt's and Dunlop's amusement. On the same trip, company engineer Samuel Strickland rode ahead and, out of sight, howled like a wolf. Dunlop then galloped ahead of Smith, and also out of sight, answered the howls! Smith was so scared that he "wheeled into the woods," only to be swept off his horse by branches. "It took three glasses of whiskey punch to work a restoration," say the Lizars. Said Dunlop before heading out, "The Accountant has taken it into his head to accompany us, and as he has never been in the bush before, won't we put him through his facings!" Robina and Kathleen Macfarlane Lizars, *In the Days of the Canada Company* (Toronto: William Briggs, 1896) 74.

71. AO, Canada Company Papers, *Minutes of the Court of Directors*, December 18, 1828.

72. AO, Canada Company Papers, *Minutes of Committees*, Committee of Correspondence, January 2, 1829.

73. AO, Canada Company Papers, *Minutes of Committees*, Committee of Correspondence, January 22, 1829.

74. Ibid.

75. AO, Canada Company Papers, *Minutes of Committees*, Committee of Correspondence, December 28, 1827.

76. AO, Canada Company Papers, *Minutes of Committees*, Committee of Treasury and Accounts, December 29, 1828.

77. Ibid.

78. AO, Canada Company Papers, *Canada Company Correspondence*, M.R. Griffen to Court of Directors, February 25, 1828.

79. AO, Canada Company Papers, *Minutes of Committees*, Committee of Treasury and Accounts, December 29, 1828.

80. AO, Canada Company Papers, *Canada Company Correspondence*, Smith to Court of Directors, January 15, 1829.

81. AO, Canada Company Papers, *Commissioners Reports*, Galt to Court of Directors, January 31, 1829.

82. £1 sterling equalled $4.44 Canadian; £1 "currency" equalled $4.00 Canadian equalled $4.00 U.S. See "Money in Canada," page 12.

83. Ibid, Directors to Galt, April 9, 1829.

84. Ibid, Fullarton to Galt, February 12, 1829.

85. Ibid.

86. See Hartwell Bowsfield, "Maitland, Sir Peregrine," in *Dictionary of Canadian Biography*, *Volume viii* (Toronto: University of Toronto Press, 1985) 596-605.

87. See Alan Wilson, "Colborne, Sir John," in *Dictionary of Canadian Biography*, *Volume ix* (Toronto: University of Toronto Press, 1976) 137-144.

88. AO, Canada Company Papers, *Commissioners Reports*, Fullarton to Galt, February 12, 1829.

89. Ibid.

90. John Galt, *The Autobiography of John Galt*, II, 130.

91. AO, Canada Company Papers, *Commissioners Reports*, Galt to Directors, January 31, 1829.

92. Ibid, Downie to Galt, March 19, 1829.

93. Ibid.

94. Ibid.

95. W.J. Van Veen, "Van Egmond, Anthony Jacob William Gysbert," in *Dictionary of Canadian Biography*, *Volume vii* (Toronto: University of Toronto Press, 1988) 882-83. Born Antonij Jacobi Willem Gijben, Van Egmond concealed his original identity. Claiming to be a descendent of the counts Van Egmond, he took on that venerable name after fleeing Holland for Germany because of some involvement in criminal activity when he was in his 20s. Van Veen reports that Van Egmond did not have the glorious military career in Napoleon's army that some believed he had had, but indicates he was probably involved on the fringes of military activity in the merchandising of supplies and transportation. Whatever his prior history, Van Egmond went to Pennsylvania in 1819 and bought land which was subsequently seized for taxes in 1826. He then moved to Upper Canada, befriended John Galt and made a significant contribution in the Huron Tract by building main roads and establishing inns. In the early 1830s, he became critical of the Canada Company in part because he was paid for his work primarily in land, not cash, and because the company, in his view, had not done enough to support settlers and their infrastructure needs. Van Egmond supported Mackenzie in his uprising in December 1837, was jailed and died from illness in January 1838. He had five sons and three daughters. In 1835, he had run unsuccessfully for the legislature but was defeated by Captain Robert Dunlop, Tiger Dunlop's brother. See James Scott, *The Settlement of Huron County* (Toronto: The Ryerson Press, 1966) 27-31, for a quite different account of "Colonel Anthony Van Egmond" and his lineage/military background *et al* as a "direct descendant of Count Van Egmond."

96. AO, Canada Company Papers, Commissioners Reports, Downie to Galt, March 19, 1829.
97. Ibid.
98. Ibid.
99. Ibid, Ellice to Galt, April 9, 1829.
100. Ibid.
101. Ibid.
102. Ibid.
103. Ibid.
104. John Galt, *The Autobiography of John Galt*, II, 136-137.
105. AO, Canada Company Papers, *Commissioners Reports*, Ellice to Galt, April 9, 1829.
106. Ibid, Galt to Dunlop, February 17, 1829.
107. As reported in *Commissioners Reports*, Galt to Directors, April 5, 1829.

Chapter 3: A Crucial Year

1. G.M. Craig, *Upper Canada: The Formative Years, 1784-1841* (Toronto: McClelland & Stewart, 1963) 138.
2. John Hodgson and William Williams resigned early in 1829, while Simon McGillivray resigned in September over a remuneration issue in September of that year. Williams' letter of resignation is instructive. He states, in part: "You know I have long felt uncomfortable at retaining my situation in the Direction without being able to attend any of the duties. Nor could I have been induced to keep it, but that I have been induced to keep it, but that it was considered desirable to reduce the number of the Directors and that no emolument was attached to it. On the contrary to me it was an expense as I have several times come to London on purpose to attend Meetings. I must therefore beg to resign; at the same time it will not be without personal regret to myself. For under different circumstances, few things could have given me more pleasure than cooperating with the Directors in promoting as far as possible the prosperity of the Concern. Pray assure them individually of my continued regard and best wishes." AO, Canada Company Papers, *Minutes of Committees*, Committee of Correspondence, February 19, 1829.
3. For a comprehensive account of Edward Ellice's life, see James M. Colthart, "Ellice, Edward," in *Dictionary of Canadian Biography, Volume ix* (Toronto: University of Toronto Press, 1976) 233-39 and *Dictionary of National Biography: From Earliest Times to 1900, Volume vi* (London: printed since 1917), 664-665.
4. For an overview of the life of Thomas Mercer Jones, see Roger D. Hall, "Jones, Thomas Mercer," in *Dictionary of Canadian Biography, Volume ix* (Toronto: University of Toronto Press, 1976) 415-17.
5. AO, Canada Company Papers, *Minutes of Committees*, Committee of Correspondence, January 8, 1829.
6. Ibid.
7. Ibid.

8. For an overview of the life of George Markland, see Robert J. Burns, "Markland, George Herchmer," in *Dictionary of Canadian Biography, Volume ix* (Toronto: University of Toronto Press, 1976) 534-36.

9. AO, Canada Company Papers, *Minutes of Committees*, Committee of Correspondence, January 8, 1829.

10. For an overview of William Allan's life and the significant contributions he made to this country, see "In Collaboration" (authors not identified), "Allan, William," in *Dictionary of Canadian Biography, Volume viii* (Toronto: University of Toronto Press, 1985) 4-13.

11. Ibid. While a Canada Company commissioner, Allan was elected in 1834 as the first president of the Toronto Board of Trade. In 1837, he became a director and then president of the City of Toronto and Lake Huron Railway. He also continued to serve in his capacity as both a legislative councillor and an executive councillor. He retired from these positions on the union of the Canadas in 1841. He served on various special commissions including investigating persons arrested for high treason after the rebellion of 1837-38; the investigation, in 1838, of the sexual misconduct of George Markland; and, in 1839-40, inspection of the administration of public departments. After he left the Canada Company commissioner position in 1841, Allan chaired meetings of the British Constitutional Society during the 1840s, and was vice-president of the British American League in 1849. He supported John Strachan in the founding of Trinity College in Toronto two years later and was a member of its board until his death in 1853. He also sat on the board of the Church Society of the Anglican Diocese of Toronto. All the while during this period, he continued to mange his extensive land holdings and investments.

    William Allan and his family lived in an impressive mansion on Sherbourne Street south of Shuter Street. He had a vibrant and successful business career and was one of the richest men in Toronto, but he had a sad personal life. Nine of his eleven children died before they were twenty. His wife died in 1848, while his daughter died in 1850. Only one son, George William, survived him. Allan's reputation "rested on his success in business."

12. AO, Canada Company Papers, *Minutes of Committees,* Committee of Correspondence, January 22, 1829.

13. AO, Canada Company Papers, *Minutes of the Court of Directors*, January 22, 1829.

14. LAC, Canada Company Papers, 1825-1887, *Minutes of a General Court of Proprietors,* April 29, 1829, 5.

15. Ibid. A marginal note in the *Minutes* of March 31, 1829 indicates the feeling of the recorder: "What can Canadians think of a Ministry who could dole out by millions for 1/2 its value & make those pay -15/ -20 -30/- who (illegible) to civilize Canada. – Fools and greedy Dogs after Govent. [sic] had so generously given £48,000 of the purchase money for Roads etc. etc. and Swamps of 11,000 acres which when drained will be excellent Land." LAC, Canada Company Papers, 1825-1887, *Minutes of a General Court of Proprietors,* March 31, 1829.

16. LAC, Canada Company Papers, 1825-1887, *Minutes of a General Court of Proprietors,* April 29, 1829, 14.

17. AO, Canada Company Papers, *Correspondence with His Majesty's Government,* Bosanquet to Hay, May 21, 1829.

18. For an overview of the life of George Murray, see Philip Buckner, "Murray, Sir George," in *Dictionary of Canadian Biography, Volume ix* (Toronto: University of Toronto Press, 1976) 639-41. Murray was born in Ochtertyre, near Crieff, Scotland, in 1772. Educated at the University of Edinburgh, he joined the military in 1789. He quickly worked his way up through the ranks. For his role in organizing the Duke of Wellington's advance through Portugal and Spain in 1812-14, he was awarded a Portuguese knighthood in 1813. On April 4, 1815, he was appointed commander of the forces in British North America and provisional lieutenant governor of Upper Canada. However, on renewal of hostilities in Europe, he quickly returned there in late May but missed the Battle of Waterloo, much to his annoyance. He was subsequently made chief of staff of the army of occupation and "lived in splendour in Paris and Cambrai." From 1820, he lived in England in an illicit relationship with Lady Louisa Erskine, the wife of British Lieutenant General Sir James Erskine. When Erskine died in 1825, Murray married his widow, but this episode hurt him socially. In November of 1819, the allied army which had been put together to defeat Napoleon was disbanded and Murray was appointed governor of the Royal Military College. In 1824, he was elected to parliament and that year he was also selected to be lieutenant general of the Board of Ordinance. He went on to command the forces in Ireland from 1825 until 1828 when Wellington, who had become prime minister in January, 1828, picked him to be colonial secretary.

   Murray's appointment was popularly received. Archdeacon Strachan was especially pleased, having met him during his brief sojourn in Canada. Murray, however, turned out to be a weak colonial secretary. Furthermore, he did not enjoy being in this position nor did he do well in either the House of Commons or around the Cabinet table. He had difficulty making decisions. He vacillated on a number of important issues pertaining to Canada, including the long-standing Clergy Reserves question and the implementation of reforms concerning the administration of the Canadas. By 1830, Wellington had no choice but to demote him to a less onerous position.

   In 1825, Murray became a lieutenant general, and a general in 1841. While he had a solid reputation in the military, and was the most decorated soldier of his day, he was one of the weakest colonial secretaries of the 19th century. If he vacillated on most matters political, Murray certainly didn't do so when it came to the Canada Company.

19. LAC, Canada Company Papers, 1825-1887, *Minutes of a General Court of Proprietors,* April 29, 1829, 5.

20. Ibid, 4-15.

21. AO, Canada Company Papers, *Correspondence with His Majesty's Government,* Bosanquet to Hay, May 21, 1829.

22. Ibid.

23. Ibid.

24. AO, Canada Company Papers, *Correspondence with His Majesty's Government,* Price to Hay, December 12, 1829.

25. Ibid, December 26, 1829.

Chapter 4: A Decade of Allan and Jones

1. "In Collaboration," "Allan, William," in *Dictionary of Canadian Biography, Volume viii* (Toronto: University of Toronto Press, 1985) 12.

2. AO, Canada Company Papers, *Commissioners Reports*, Bosanquet to Allan, January 27, 1829.

3. See Robina and Kathleen Macfarlane Lizars, *In The Days of the Canada Company* (Toronto: William Briggs, 1896), Chapter vii, "The Colborne Clique," 117-150. "The Hyndmans, Lizars, Kippens, Lawsons, Clarkes, John Galt jun, and a host of others were of the Clique; while some other among the English contingent of gentlemen emigrants, though not of the Clique, were anti-Canada Company. The Clique had friends on both sides of the [Maitland] river. Dissatisfaction began early, chiefly from disappointment at findings not as the Company's maps and illustrations in London and Edinburgh led purchasers to expect; in many cases, reality did not tally with the scenes conjured up by imagination. Froude [possibly a reference to Sigmund Freud] says somewhere that when the wise and good are divided in opinion the truth is generally found to be divided, too."

    While the Clique may have been anti-company, they still happily partook in company happenings. For instance, at the launch of the second company boat, the *Minnesetung,* in July, 1834, Daniel Lizars' daughter was front and centre at the christening ceremony. Pretty Helen Lizars stood by John Galt the younger on the deck; Doctor and Captain Dunlop, red-shirted as backwoodsmen, and the Colborne Clique and the Canada Company men, all expectant and all more or less picturesque in appearance stood about. The young girl broke a bottle of wine as she pronounced the word '*Minnesetung'* in a sweet treble voice, and the vessel was launched."

4. Ibid.

5. John Galt Jr. was the son of Elizabeth Tilloch and the Canada Company's John Galt originally of Irvine, Scotland. John Jr. was born in Great Britain in 1814 and had two brothers – Thomas (born in 1815) and Alexander (born in 1817 in London). During this period, John Galt Sr. and his family moved more than once between London and Scotland and it has not been possible to ascertain exactly where John Jr. and Thomas were born. When John Sr. moved to Canada in 1826 to take up his duties as Canada Company commissioner, Elizabeth and their sons remained in Scotland where the boys attended Reading School in Eskgrove near Edinburgh (the Reverend Dr. Richard Valpy, headmaster). In the spring of 1828, John's wife and sons arrived in Upper Canada via New York to join him. They proceeded to Burlington Bay where the Galts lived as a family until the end of the summer. The boys were then sent off to school in Chambly, Lower Canada (Quebec), while the John and Elizabeth moved to Guelph to take up what they thought would be permanent residency in the Priory. However, John was to leave for England in March of 1829 in order to meet with company directors to discuss the situation in Canada and his future. He arrived in London in May. When Mrs. Galt realized her husband would not be returning to Upper Canada, she then moved to Chambly to be near the boys and was on her own for the best part of the year.

Following Galt's arrival in England, he not only had to meet with company directors, but his creditors closed in on him as well. They included the Reverend Dr. Valpy who had not been paid for tutoring the Galt boys and whose lawyer pressed charges. Galt ended up in debtor's prison from July 15 to November 10, 1829, where he wrote prodigiously. By the summer of 1830, Galt's family had returned to Great Britain (probably London).

In 1833, John Jr. returned to Upper Canada with his brother Thomas, while Alexander followed in 1834. Settling in Colborne Township just north of Goderich, John Jr.'s involvement in political issues included siding with Tiger Dunlop in the disputed election of 1841 which pitted Dunlop (who had fallen out with the company by then) against Canada Company man Captain James McGill Strachan (he was Bishop Strachan's son and Jones's brother-in-law). John Jr. died in 1860.

Thomas worked for the Canada Company for six years, then took up the study of law and ultimately became Chief Justice of the Supreme Court of Ontario. Alexander took up a position in the British American Land Company in Sherbrooke Quebec which was opening land for settlement in the Eastern Townships (John Galt Sr. had been able to resurrect himself in London and became secretary of the company for nine months in 1832 before ill health caused him to resign in December of that year). While in the employ of the British American Land Company, Alexander began promoting railways and ultimately became president of the St. Lawrence and Atlantic Railway. He was a member of the legislature of Canada (1849-50; 1853-67), a "Father of Confederation" and Canada's first High Commissioner to London. He was also knighted. See Roger Hall and Nick Whistler, "Galt, John," in *Dictionary of Canadian Biography, Volume vii* (Toronto: University of Toronto Press, 1988) 335-340; H.B. Timothy, *The Galts: A Canadian Odyssey: John Galt 1779-1839* (Toronto: McClelland & Stewart, 1977); Margaret E. McCallum, "Galt, Sir Alexander Tilloch," in *The Canadian Encyclopedia* (3 volumes) (Edmonton: Hurtig Publishers, 1985) 716-17.

6. For information on the Buffalo and Lake Huron Railway, see http://www.globalserve.net/~robkath/railcn.htm (Buffalo and Lake Huron Railway), downloaded June 2004.

7. AO, Canada Company Papers, *Minutes of the Court of Directors*, Instructions, January 22, 1829.

8. Ibid.

9. Ibid.

10. Ibid.

11. Ibid.

12. Ibid.

13. Ibid.

14. Ibid.

15. Ibid.

16. AO, Canada Company Papers, *Commissioners Reports*, Ellice to Commissioners, June 12, 1829.

17. AO, Canada Company Papers, *Minutes of Committees*, Committee of Correspondence, May 28, 1829.

18. AO, Canada Company Papers, *Commissioners Reports*, Allan to Directors, September 15, 1829.
19. Ibid, Price to Commissioners, October 22, 1829.
20. Ibid, January 30, 1830.
21. Ibid. January 22, 1830.
22. The Priory probably cost £1,000 to build, as Galt had been paying £100 a year rent and had always said that a building should rent for 10% of its value.
23. AO, Canada Company Papers, *Commissioners Reports*, Jones to Directors, May 6, 1829.
24. The first step leading to the ultimate purchase by the settler was for the company to obtain land grants from the government as specified in the agreement. Galt had not kept the directors informed and so they had no idea whether the grants had in fact been made.
25. AO, Canada Company Papers, *Commissioners Reports*, Jones to Directors, May 6, 1829.
26. Ibid, Allan to Directors, July 7, 1829.
27. Ibid, Jones to Directors, June 1, 1829.
28. Ibid, McGillivray to Commissioners, July 23, 1829.
29. Ibid, Jones to Directors, June 1, 1829.
30. Ibid.
31. Ibid, June 1, 1829.
32. Jones was a bachelor until 1832 when, at age 37, he married Archdeacon John Strachan's daughter, Elizabeth Mary, in Toronto. The next year he built a fine house there for his bride at York and Front streets on property severed from the Archdeacon's land. Although based in Toronto until moving with his family to Goderich in 1840, Jones spent a considerable amount of time in Goderich each year overseeing Canada Company business in the Huron Tract and travelling the territory. Initially he lived in the Canada Company headquarters building The Castle (on the current Harbour Park property) which had been built by Tiger Dunlop. He then built his own house high on the lakeside bank, on the south side of the harbour overlooking the lake and the mouth of the Maitland River. He held on to that property upon moving to Goderich, but sold it to "Her Majesty" in the 1849 for construction of the present lighthouse.

    Jones was considered a cad by some, while others reported him to be a "humane and talented commissioner." He reportedly kept a mistress in Goderich whom he handed over to Charles Prior upon moving there (Galt had sent Prior to Goderich from Guelph in 1828 as superintendent of works for the company). That arrangement lasted for about a year when Prior dropped her.

    Annette Stewart wrote of Jones: "Jones was an Irishman of some business ability, but was known in Huron chiefly for the bad company which he often kept and for his low cunning which he displayed on occasion. For some years he employed as head agent in Goderich a certain John Longworth, an Irish ex-survivor of the Peninsula War, who was rough and domineering and had been convicted of bigamy. Jones and Longworth favored the Irish and had a large group of them in and about Goderich."

    While Longworth and Prior were certainly not the best of ambassadors for the company, there was no way of the directors really knowing what was going on.

Furthermore, Jones' tight connection with Family Compact members would have stood him in good stead despite cost overruns on virtually every project initiated by him. Prior was viewed by settlers as being rude and tyrannical, yet with peers and superiors he was known to be very suave. He was fired in 1836 for embezzling the company, but that did not stop Jones from maintaining his friendship with him. Surprisingly, Prior was subsequently appointed a justice of the peace in Goderich. Given his instructions of 1829 before leaving for Canada, one is left to wonder whether Jones ever placed a surety bond on Prior. See Annette Stewart, "The 1841 Election of Dr. William Dunlop as Member of Parliament for Huron County," *Ontario History, Vol. xxxix*, Ontario Historical Society, Toronto (published by the Society, 1947), 51; James Scott, *The Settlement of Huron County* (Toronto: Ryerson Press, 1966), 66–71.

33. AO, Canada Company Papers, *Commissioners Reports*, Jones to Directors, July 7, 1829.

34. Despite Galt's significant expenditures at Guelph, a gristmill had not been built, although authorized by the directors.

35. AO, Canada Company Papers, *Commissioners Reports*, Easthope to Commissioners, September 3, 1829.

36. A review of Canada Company records indicates that meetings of the General Court of Proprietors and Court of Directors were invariably held in the London Tavern rather than in the Canada Company offices. For their July 12, 1826, meeting, the company was billed £6.17.8 by the tavern for the use of a meeting room. Undoubtedly, this tavern was close to the company's offices on Bishopgate Street. AO, Canada Company Papers, *Minutes of the Court of Directors*, November 17, 1826.

37. Ibid, Jones to Directors, November 9, 1829.

38. Ibid, Allan to Directors, September 7, 1829.

39. Ibid.

40. As of September 1, 1829, 75,281 acres of land had been sold for a total £39,499. Of this amount, £12,771 had been paid in cash, with the balance to be paid in installments (£26,728) earning the company six per cent per annum.

41. AO Canada Company Papers, *Commissioners Reports*, Allan to Simon McGillivray, September 7, 1829.

42. AO Canada Company Papers, *Minutes of the Court of Directors*, September 3, 1829.

43. Ibid.

44. Ibid, Allan to Directors, October 8, 1829.

45. See Thomas Smith, *An Address to the Shareholders of the Canada Company*, (London: 1829).

46. AO, Canada Company Papers, *Commissioners Reports*, Allan to Directors, October 8, 1829.

47. Ibid, Jones to Directors, October 19, 1829.

48. Ibid, Allan to Directors, October 8, 1832.

49. Ibid, October 8, 1829.

50. See John C. Weaver: "Cull, James," in *Dictionary of Canadian Biography, Volume vii* (Toronto: University of Toronto Press, 1988) 221-23, for a short account of Canada Company emigration agent William Cattermole.

51. Ibid, 221-23.

52. Initially, free passage was provided on payment of the first instalment at either Quebec City or Montreal. In 1832, however, it was decided that for the 1833 season, free passage would be limited to those who had purchased one hundred acres in the Huron Tract. The emigrant would pay the conveyance cost up front, and would then be allowed to deduct it from the second instalment.

53. AO, Canada Company Papers, *Commissioners Reports*, Price to Commissioners, February 13, 1830.

54. Ibid, November 18, 1830.

55. Tiger Dunlop's brother, former Royal Navy Captain Robert Graham Dunlop, was the captain of the *Menesetunk* for a relatively short period. He had arrived in Upper Canada in 1833 and had shortly thereafter been appointed a commissioner of the "Court of Requests." The company obviously needed look no further than Dunlop for a captain for their boat, but the relationship was not to last long. Some accounts indicate that he was relieved of his command in 1836 for what the company viewed as extravagance and incompetence.

56. Baron de Tuyll had bought this site 12 miles south of Goderich on the advice of hydrographic surveyor Henry Wolsey Bayfield whom he had retained to select a place build a community. (Bayfield had made an extensive survey of Lake Huron and the rivers running into it about 1820.) However, when the Baron died in 1835, little progress had been made and his dream died with him. "... the work seems to have been ill-advised; for in 1836 – the year of the fat, dark little Baron's death – logs were lying rotting, the buildings consisted only of his store and a few huts, ..." Robina and Kathleen Macfarlane Lizars, *In the Days of the Canada Company* (Toronto: William Briggs, 1896) 106.

57. For an overview of Henry Wolsey Bayfield's life and work, see Ruth McKenzie, "Bayfield, Henry Wolsey," in *Dictionary of Canadian Biography, Volume xi* (Toronto: University of Toronto Press, 1982) 54-57.

58. University of Guelph Library, Arts Division, *Report of the Court of Directors of the Canada Company to the Proprietors,* London: March 23, 1832.

59. AO, Canada Company Papers, *Commissioners Reports*, unsigned letter to Commissioners, March 29, 1832.

60. University of Guelph Library, Arts Division, *Report of the Court of Directors of the Canada Company to the Proprietors,* London: March 23, 1832.

61. AO, Canada Company Papers, *Commissioners Reports*, Price to Commissioners, February 12, 1831.

62. Ibid, Perry to Commissioners, August 30, 1832.

63. J.J. Talman, "Merritt, William Hamilton," in *Dictionary of Canadian Biography, Volume ix* (Toronto: University of Toronto Press, 1976) 546.

64. AO, Canada Company Papers, *Commissioners Reports*, Price to Commissioners, June 28, 1832.

65. Ibid, December 19, 1833.

66. University of Guelph, Arts Division Library, Anonymous *To The Shareholders of the Canada Company* (London: 1832).

67. Ibid.

68. York was named and laid out as a town in 1793 by Upper Canada's first governor, John Graves Simcoe. It was renamed "Toronto" (an Aboriginal name meaning "meeting place") in 1834. By then it had a population of some 9,000 and had an elected civic government. Its first mayor, who only lasted one year, was, surprisingly, William Lyon Mackenzie. See J.M.S. Careless, "Toronto," in *The Canadian Encyclopedia* (3 volumes) (Edmonton: Hurtig Publishers, 1985) 1831-1834.

69. AO, *Correspondent and Advocate* (Toronto), December 11, 1834.

70. AO, *Patriot and Farmer's Monitor* (Toronto), June 12, 1835.

71. The Huron Tract lands were subjected to taxation as soon as surveyed and paid for. The Crown Reserves, by right of pre-emption, were not taxed until applied for by the company. The company at this point was paying approximately £1,800 annually in taxes on their unsold land in the Tract.

72. AO, *Correspondent and Advocate* (Toronto), May 7, 1835, Report from the Committee of Grievances, Evidence of James Wilson, M.P.P.

73. AO, *Correspondent and Advocate* (Toronto), August 20, 1835.

74. 18 GEO. 11. C 12 – Acts of Parliament are designated this way. "18 GEO" means the act was passed in the 18th year of the reign of George 11 while "C 2" refers to the Parliamentary number of the Act.

75. AO, "Politics in Upper Canada" in *Correspondent and Advocate* (Toronto), August 20, 1835.

76. Fred Landon, *Lake Huron* (New York: The Bobbs-Merrill Co., 1944) 123.

77. AO, Canada Company Papers, *Minutes of Committees*, Committee of Correspondence, February 19, 1835.

78. AO, Canada Company Papers, *Commissioners Reports*, Perry to Commissioners, January 21, 1836.

79. Robina and Kathleen Macfarlane Lizars, *In the Days of the Canada Company* (Toronto: William Briggs, 1896) 120.

80. G.H. Needler, *Colonel Anthony Van Egmond* (Toronto: Burns and MacEachern, 1956) 22-23.

81. AO, *Correspondent and Advocate* (Toronto), January 11, 1835.

82. AO, Canada Company Papers, *Other Letters of the Board of Directors*, Perry to Van Egmond, April 9, 1835.

83. University of Guelph, Arts Division Library, *Report of the Court of Directors of the Canada Company to the Proprietors – Supplement*, London, March 24, 1836.

84. AO, Canada Company Papers, *Commissioners Reports*, as reported in a letter from Perry to Commissioners, August 28, 1835.

85. University of Guelph, Arts Division Library, *Report of the Court of Directors of the Canada Company to the Proprietors* – Supplement, London, March 24, 1830.

86. AO, Canada Company Papers, *Commissioners Reports*, Perry to Commissioners, May 22, 1834.

87. University of Guelph, Arts Division Library, *Report of the Court of Directors of the Canada Company to the Proprietors* – Supplement, London, March 24, 1830.

88. AO, Crown Land Papers, *Correspondence*, Hugh Black to the Honourable R.B. Sullivan, October 22, 1836.

89. Ibid. Black says that his information was supplied by a man of intelligence from the London District, "... he tells truths we could never know from Newspaper puffs."

90. AO, Crown Land Papers, *Correspondence*, Black to Sullivan, October 22, 1836.

91. Ibid.

92. W.H. Graham, *The Tiger of Canada West* (Toronto: Clarke, Irwin, 1962) 155.

93. Ibid, 166.

94. AO, Canada Company Papers, *Correspondence with Commissioners*, Perry to Commissioners, June 1, 1837.

95. Ibid, July 14, 1836.

96. Ibid, June 1, 1837.

97. While there was much enthusiasm amongst promoters to construct the railway which would ultimately run from Niagara Falls to Windsor through Hamilton and London, it would be 1849 before ground would be broken and January 1854 before the railway would be open for business. Initially incorporated in May 1834 as the London and Gore Railroad Company, it became the Great Western Railroad Company in 1845 and the Great Western Railway in 1853. See Peter Baskerville, "Great Western Railway," in *The Canadian Encyclopedia* (3 volumes) (Edmonton: Hurtig Publishers, 1995) 772.

98. AO, Canada Company Papers, *Correspondence with Commissioners*, Perry to Commissioners, June 1, 1837.

99. Ibid.

100. Ibid.

101. Ibid.

102. Ibid. Company costs in pounds sterling that year were computed by the company as follows: payment to government – £20,000; interest to proprietors – £14,000; expense of establishment in Canada – £5,500; expense of establishment in England – £2,500; Total: £42,000.

103. AO, Canada Company Papers, *Correspondence with Commissioners*, Perry to Commissioners, November 16, 1837.

104. Ibid.

105. Ibid.

106. Ibid.

107. The agents in Upper Canada seemed to be on a small salary, plus a commission of 5% for land sold.

108. PAO, Canada Company Papers, *Commissioners Reports*, Perry to Commissioners, July 27, 1837.

109. AO. Canada Company Papers, *Correspondence with Commissioners*, Perry to Commissioners, November 16, 1837.

110. The commissioners would have appreciated this support because William Lyon Mackenzie was continuing to create trouble. The following message addressed to the farmers of York County was distributed in a broadside in late November – just days before Mackenzie's followers marched on Toronto. It included, among other comments about tariffs, freedom, corrupt government etc:

> CANADIANS! It is the design of the Friends of Liberty to give several hundred acres to every Volunteer – to root up the unlawful Canada Company, and give *free deeds* to all settlers who live on their lands …so that the yeomanry may feel independent, and be able to improve the country, instead of sending the fruit of their labour to foreign lands.

Quoted in Robert Weaver and William Toye eds., *The Oxford Anthology of Canadian Literature* (Toronto: Oxford University Press, 1973) 297. Taken from "The Selected Writings of William Lyon Mackenzie, 1824-1837" by Margaret Fairley, 1960.

111. See Graham, 173-181.

112. AO, Canada Company Papers, *Canada Company Correspondence*, Perry to Hyndman, March 29, 1838.

113. Ibid, Perry to R.G. Dunlop, March 29, 1838.

114. AO, Canada Company Papers *Commissioners Reports*, Perry to Commissioners, March 8, 1838.

115. Ibid, March 29, 1838.

116. University of Guelph, Arts Division Library, *Report of the Court of Directors of the Canada Company to the Proprietors*, London, December 31, 1839.

117. See Reports of the Court of Directors of the Canada Company to the Proprietors for the respective years.

118. AO, Canada Company Papers, *Correspondence with Commissioners*, Perry to Commissioners, January 3, 1839.

119. Ibid, "Cost of Sundry Undertakings in the Huron Tract, with Interest to 31st December 1839," October 15, 1840.

120. AO, Canada Company Papers, *Commissioners Reports*, Perry to Commissioners, July 10, 1834.

121. See P.E. Lewis, "Faction in the Goderich Area, Huron County," unpublished M.A. thesis, University of Western Ontario, 1967, Chapter III, for the politics surrounding the building of the gaol.

122. AO, Canada Company Papers, *Correspondence with Commissioners*, Perry to Commissioners, January 3, 1839.

123. Ibid.

124. Ibid.

125. Ibid.

126. Ibid.

127. AO, Canada Company Papers, *Minutes of the Court of Directors*, November 25, 1830.

128. AO, Canada Company Papers, *Correspondence with Commissioners*, Perry to Commissioners, March 3, 1836.

129. Thomas Mercer Jones and William Allan were quite well paid for their services. In 1838, Allan was earning a salary of £600 sterling for his less than full time work on behalf of the company; Jones was earning £750 sterling in salary, £100 currency for house rental and £300 currency for travelling expenses: as reported in AO, Canada Company Papers, *Correspondence with Commissioners*, Perry to Commissioners, April

12, 1838. But compare that to John Galt's salary in 1827 of £1000 sterling, plus allowance of £500 currency and 5% commission for actual monies received on sales and interest paid! Obviously, the directors were unaware of the going rate for salaries/allowances in 1827.

130. "In Collaboration" (authors not identified), "Allan, William," in *Dictionary of Canadian Biography, Volume viii* (Toronto: University of Toronto Press, 1985) 8.

131. Ibid, February 16, 1837. AO, Canada Company Papers, *Correspondence with Commissioners*, Perry to Commissioners, February 16, 1837.

132. AO, Strachan Papers, *Strachan Letter Book, 1827-1839,* 243.

133. AO, Canada Company Papers, *Correspondence with Commissioners*, Perry to Commissioners, February 16, 1837. This was undoubtedly an initiative of William Allan who, upon leaving the presidency of the Bank of Upper Canada, became a shareholder in the British-based Bank of British North America in 1836. The directors do not appear to have realized that Jones was in fact spending a significant amount of time in Goderich in particular, if not the Huron Tract in general.

134. AO, Perry to Commissioners, May 2, 1839.

135. B.W. Connolly, S.J., "After 100 Years – A Jesuit Seminary," (Archives, Cathedral Church of St. James, Toronto), an article in "Loretto Rainbow," January, 1946.

136. Alan Wilson, "Widder, Frederick," in *Dictionary of Canadian Biography, Volume ix* (Toronto: University of Toronto Press, 1976) 836-38.

137. Ibid. The house which became known as "Lyndhurst" when the Widders acquired it was built by Robert Sympson Jameson in 1837. It was less than a mile east of Old Fort York along the shores of Lake Ontario on five acres of an old military reserve, the Government Common. A lawyer, Jameson had married Anna Murphy in 1825. In 1833, he was sent to Upper Canada as attorney general. He and his wife had a strained relationship (they had been engaged five years before their marriage and had broken it off), and so Anna decided to take up a governess position in Germany when her husband took up his position in Canada. A formal demand from Robert in 1836 resulted in Anna leaving Germany for New York and then proceeding alone to Toronto in late fall where she received a frosty reception from her husband. In the meantime, Robert had built a home for them and they moved into it early in 1837. At that point, it was a smallish two-storey house on land that would now be bounded by Front Street, Spadina Avenue, Queen Street and Wellington Street.

Anna never liked Toronto and viewed her husband as being "cold, aloof and dictatorial" – and he drank too much, according to Anna. In the spring of that year after having lived in the house for only four months, she left her husband and Toronto to tour the western portion of the province. She reached home in August, negotiated with Robert for a £300 yearly allowance – and left for Europe. She never saw him or the house again. Anna spent the rest of her life writing and travelling in England and on the continent. She wrote 22 books during this period, one being *Winter Studies and Summer Rambles in Canada*. Robert Jameson now lived alone again. He subsequently gave up the house in 1844, having handed over the property, to the degree possible, to Frederick Widder without Anna's consent on payment of £2,000. The real estate deal was formally concluded with Widder some 2 1/2 years later. The Widders named the property "Lyndhurst" and enlarged it quite considerably. Beside the main stairway Frederick had a stained glass window installed with the motto "Nusquam Meta Mihi" – "Nothing Daunts Me" – very fitting in view of his approach to life and work.

The *Toronto Evening Telegram*, in a retrospective article of the early 1880s, reported that the Widders entertained well and were most hospitable. "Their drawing room was the centre of social activities from the mid-forties till the early sixties. The older generation yet living will remember the balls, the dinner parties, and the other social events that made Lyndhurst so popular." Both Frederick and his wife Elizabeth were connected to royalty (Austro-Bavarian and English respectively) and so their house often drew visitors of high rank from abroad, including Edward VII, then Prince of Wales, in 1860. In 1861, the Widders went to England for a visit. Frederick and his daughter Jane returned in 1862. They did not stay in Lyndhurst, but at a hotel. In 1863, Mrs. Widder returned to Canada and went to Montreal to live with her daughter where she died the next year. Mr. Widder resigned from the company for reasons of ill health in 1864 (he was practically an invalid by this stage) and died in February, 1865 in Montreal on his way back to England. By this time, the house was flanked by rail lines on the south (the first line having opened in 1853) and railway shops and round-houses – but it still had lovely gardens and was one of the finest homes in Toronto.

Following the death of the Widders, Lyndhurst was leased for short time to a Mrs. Gordon and her son and daughter while her husband tried to buy it. He was unsuccessful and, in 1867, it was acquired by the Roman Catholic Church. It became Loretto Abbey, a school for girls. It was enlarged on three occasions. In 1928, the school was closed, a new Loretto Abbey having been built in Armour Heights in North Toronto. In 1930, the property was turned over to the Society of Jesus and became a Jesuit Seminary. It was now surrounded by factories and offices on the east, north and west, with the railways on the south. The Jesuits kept it until 1962 when it was sold and torn down. Part of the Lyndhurst estate is now the site of the *Globe and Mail* newspaper. See B.W. Connolly, reprint of article from the *Loretto Rainbow*, January 1946.

Chapter 5: A Stormy Fourteen Years

1. From Archives, Cathedral Church of St. James, Toronto; "After 100 Years – A Jesuit Seminary." Reprint from *Loretto Rainbow*, January 1946.
2. English-born and Eton-educated, Lord Durham (John George Lambton) was elected to the House of Commons in 1813 as a liberal Whig and raised to the House of Lords in 1828. Suffering from poor health, he resigned from cabinet in 1833 after having made a significant contribution to the drafting of the Reform Bill of 1832. He was named ambassador to Russia from 1833 to 1837, and was then persuaded to become governor general and high commissioner to British North America with responsibility for preparing a report on the rebellions of 1837 and 1838. He arrived in Montreal in May 1838 but resigned six months later over an illegal ordinance dispute. In January 1839, and now back in London, he completed his *Report on the Affairs of British North*

America. Philip A. Buckner, "Durham, John George Lambton, 1st Earl of," in *The Canadian Encyclopedia* (3 volumes) (Edmonton: Hurtig Publishers, 1985) 525. See also David Mills, "Durham Report," in *The Canadian Encyclopedia* (3 volumes) (Edmonton: Hurtig Publishers, 1985) 525, 526.

3. The tax dispute centred on the Huron District Council attempt to levy a tax on all unsold or "wild" land at the same rate as land under cultivation. Since most of the "wild land" in the District was owned by the company, the company complained it was being unfairly penalized.

4. AO, Canada Company Papers, *Correspondence with Commissioners*, Perry to Commissioners, October 15, 1840, "Cost of Sundry Undertakings in the Huron Tract, with Interest to 31st December, 1839."

5. The Lizars' describe the building at the top of the harbour hill (the present-day "Park House," minus the "steep French roof") as follows: "The new Canada Company building, with its steep French roof, dormered third story windows, plastered walls and low ceilinged rooms, now became a castle indeed. But it would not be long before "Mrs. Thomas Mercer Jones, née Elizabeth Mary Strachan, only and very lovely daughter of his same grace, Primate of Western Canada" would find fault with the home. It did not meet the lady's approval. The entrance was soon changed, and within the mansion the stairway was reversed and many alterations made to suit her judgment and taste." Robina and Kathleen Macfarlane Lizars, *In the Days of the Canada Company* (Toronto: William Briggs, 1896) 316-17.

6. AO, Canada Company Papers, *Correspondence with Commissioners*, Perry to Commissioners, October 17, 1839.

7. Ibid.

8. W.H. Graham, *The Tiger of Canada West* (Toronto: Clarke Irwin, 1962) 195. The Misses Lizars describe how the Jones' entertained: "One night, the Canada Company building was lit from garret to cellar, and the commissioner and his wife welcomed their entire circle of acquaintance. They danced in the large dining-room on the right of the hall; the library opposite was a resting place and upstairs in the drawing room, with its two bright fires and heavy yellow damask and bullion fringe, stood the little Queen of all, in white watered silk and pearl ornaments in her black hair and on her white neck and arms. ... The big chandeliers, sconces on the wall and candelabra, held many wax-lights which shone on very happy faces. But Jimmy Collins and his fiddle were not there. Music of a more advanced kind from London took his place." Robina and Kathleen Macfarlane Lizars, *In the Days of the Canada Company,* 328-29.

9. T.M. Jones to the editor, *Toronto Patriot,* March 24, 1840.

10. AO, Canada Company Papers, *Correspondence with Commissioners*, Perry to Commissioners, December 17, 1840.

11. Ibid, July 16, 1840.

12. Ibid.

13. AO, 1840 Miscellaneous, *Preliminary and Confidential Report of an Enquiry into the Affairs of the Canada Company Addressed to His Excellency, Major General, Sir G. Arthur K.C.B.*, dated at Toronto, Upper Canada, November 2, 1840, 13-14.

14. For an overview of the life of Jonas Jones, see Robert L. Fraser, "Jones, Jonas," in *Dictionary of Canadian Biography, Volume vii* (Toronto: University of Toronto Press, 1988) 456-61.

15. AO, 1840 Miscellaneous, Preliminary and Confidential Report of an Enquiry ... (1840) 13-14.

16. Ibid, 24.

17. Ibid, 27.

18. Ibid, 28.

19. AO, Canada Company Papers, *Correspondence with Commissioners*, Perry to Commissioners, October 15, 1840.

20. Ibid.

21. The size of William Allan's land holdings is not known precisely, but it is known that he owned land in practically every district in Upper Canada, having been involved in land speculation for 60 years. In addition, he had extensive holdings in Toronto which he had purchased in 1819 (in today's terms, it would be the area from present day Richmond Street to Bloor Street and from Sherbourne Street to about Jarvis Street). In 1841, he is said to have wanted a list complied of 20,000 acres of land which he wanted to sell. See "Allan, William," *Dictionary of Canadian Biography, Volume viii* (Toronto: University of Toronto Press, 1985), 4-13.

22. Ibid.

23. AO, Canada Company Papers, Correspondence with Commissioners, Franks to Allan, October 15, 1840.

24. Ibid, Perry to Commissioners, October 15, 1840.

25. Statute Labour was a somewhat nebulous requirement halfway between a job and volunteering which required three or four days a year from a landowner. The use of labour depended a great deal on who was chosen as overseer and on the weather. Invariably, affluent land owners would pay somebody to do the work for them.

26. AO, Canada Company Papers, Correspondence with His Majesty's Government, R.V. Smith to C. Franks, January 18, 1840.

27. AO, Canada Company Papers, *Correspondence with Commissioners*, Perry to Commissioners, July 16, 1840.

28. AO, Canada Company Papers, *Correspondence with Commissioners*, Perry to Commissioners, July 28, 1840.

29. Company Crown Reserve sales for the years 1840 and 1841 totalled 51,174 acres while Huron Tract sales for the same period equalled 26,019 acres. From W.H. Smith, *Canada: Past, Present and Future* (Toronto: publisher n.k., 1851) Vol. ii, 165.

30. AO, Canada Company Papers, *Letters to the Court of Directors from Frederick Widder* (hereinafter referred to as "Letters from Widder"), June 6, 1840.

31. Ibid.

32. Ibid, February 18, 1841. Jones's remuneration/expenses per annum were still at the level reported in 1838: Salary: £750 sterling; Housing Allowance: £100 currency; Travelling Expenses: £300 currency.

33. See Graham, 207-228.

34. AO, Canada Company Papers, *Correspondence with Commissioners*, Perry to Commissioners, May 27, 1841.

35. Ibid, July l, 1841.

36. AO, Crown Land Papers, Correspondence, Order in Council and Regulations regarding Matters connected with the Department of Crown Lands, Vol. I, Canada Company, *Copy of a Report of a Committee of the Executive Council Relative to the Huron Tract*, October 16, 1841, 101.

37. AO, Canada Company Papers, *Correspondence with Commissioners*, Perry to Commissioners, March 2, 1843.

38. Ibid, August 26, 1841.

39. University of Guelph, *Aitchison Brown Diary* (microfilm copy), 66.

40. As reported in AO, Canada Company Papers, *Correspondence with Commissioners*, Perry to Commissioners, March 10, 1842.

41. Ibid.

42. Ibid.

43. University of Guelph, Arts Division, see Reports of the Court of Directors of the Canada Company to the Proprietors for 1841 and 1842.

44. AO, Canada Company Papers, *Correspondence with Commissioners*, Perry to Commissioners, March 10, 1842.

45. Ibid, September 1, 1842. Thomas Mercer Jones would have considered William Geary to be a good risk. According to a list of eleven settlers compiled in the early 1840s, the Lizars reported that William Geary had £300 in cash, 200 acres of land (30 acres of which was cleared), two oxen, six cows, and eight young cattle. All on the list, they reported, were "frugal, industrious men, three fourths of whom had been farmers in the Old Country, and the remainder mechanics. William Geary, the first named on the list, instituted and ran the first daily stage coaches between London and Goderich; he also had the contracts of supplying timber brought from the Sauble for the much-criticized Colborne Bridge and other company works. These men were all young and had been in the country five years." Robina and Kathleen Macfarlane Lizars, *In the Days of the Canada Company* (Toronto: William Briggs, 1896) 412-13.

46. AO, Canada Company Papers, *Correspondence with Commissioners*, Perry to Commissioners, September 1, 1842.

47. Ibid.

48. While management was having its challenges with Thomas Mercer Jones, he and his wife relished the social scene in Goderich. In addition to entertaining lavishly, they were ostentatious in their daily living habits. On Sundays, they and their children who survived (Charles Mercer, born in 1838, and Strachan Graham, born in 1841), plus their servants, made quite an entry to their church (Church of England): "It must be confessed that there is a good deal of nabob and Ranee about the Commissioner's establishment. Naturally, the butterfly queen was the cynosure of all eyes, and many tales, most of them pleasing, some amusing, come down to us of the times she made her own. In the new stable-church [in the latter 1830s, a barn on West Street near the Canada Company headquarters had been converted into a place of worship for Church

of England adherents until the first brick St. George's Church was built in 1843 on land donated by the company] the Canada Company held two pews, one of which was occupied by the Jones family and one by servants. The latter always filed in first, lady's maid, nurse, two men-servants, sometimes more sometimes less of the total of eight."

On another occasion: "A commotion at the [church] door weekly resolved itself into a procession in which the Commissioner and his wife came first, followed by those who formed the perennial house party... Beautifully dressed, the Ranee knelt far into the middle of the square, and the congregation, one half admiring and the other criticizing, divided their attention between her and the mellow accents of the Rector..." Robina and Kathleen Macfarlane Lizars, *In the Days of the Canada Company* (Toronto: William Briggs, 1896) 324–25.

49. AO, Canada Company Papers, *Correspondence with Commissioners*, Perry to Commissioners, September 1, 1842.

50. Ibid.

51. Ibid, August 11, 1842.

52. Ibid, November 17, 1842.

53. Ibid, June 29, 1843.

54. According to the Lizars, Henry Ransford went to the Huron Tract in 1832. "Mr. Ransford ... was ... a most enterprising citizen and a thoroughly good immigrant. He was at daggers drawn with the Canada Company, his prejudices paving the way for one of those duels which 'might have been.' Yet he was not of the Colborne Clique, but a kind of free lance among them all, sometimes friendly, sometimes with the coolness of friendship approaching armed neutrality. He was a dashing, handsome Englishman with an equally handsome American wife; had been in a good regiment, had studied medicine in London and Paris, and was altogether one of those life-tasters with whom this period abounded." He "took up" several thousand acres in Tuckersmith Township where he built a homestead. *In the Days of the Canada Company*, 300–01.

In "later years," the Lizars report that Henry Ransford became a commissioner of the Canada Company. "It is part of a good shepherd to shear, not flay, his sheep." Ibid, 286.

55. AO, Canada Company Papers, *Correspondence with Commissioners*, Perry to Commissioners, December 4, 1843.

56. University of Western Ontario, John Galt Jr. Papers, Galt to Jones, July 15, 1844.

57. Ibid.

58. Although Thomas Mercer Jones may have been somewhat of a cad while in Goderich on his own for extended periods of time, in his early married days he obviously revered his wife, Elizabeth Mary, who was 19 years his junior. What better way to immortalize her than to have a community named in her honour, with "St."(for Saint) added for good measure! "Away off in the very inner heart of the Tract, over the stony bottom of the Thames, was a ripple which went by the name of Little Falls... As early as 1841, the Commissioner decided that there a town should be founded. He went again about 1845, accompanied by Mrs. Jones, to give the town an official name, for

the lack of such led to confusion in the giving of titles. A discussion arose at the meeting called as to the name, and she, to the rescue as usual, suggested it should be called after herself. St. Mary's it then and there became, the beginning of a thriving and picturesque town" [20 kilometres or 12 miles southwest of Stratford]. "Mrs. Jones, after laying this foundation, of a memory ... said she would donate ten pounds towards a school." Robina and Kathleen Macfarlane Lizars, *In the Days of the Canada Company*, 372. Given the times, it is most likely that this scenario would have been orchestrated and pre-arranged by Jones.

59. "A Huron," "The Canada Company. To the Editor of the British Colonist," *British Colonist*, July 1, 1840.

60. AO, Canada Company Papers, *Letters from Widder,* Widder to Directors, December 11, 1841.

61. Ibid. To hire a wagon to convey a settler, his family and effects from Hamilton to Goderich cost $10.00 or approximately £2/10 currency.

62. AO, Canada Company Papers, *Letters from Widder*, Widder to Directors, December 11, 1841.

63. AO, Canada Company Papers, *Correspondence with Commissioners*, Perry to Commissioners, January 27, 1842.

64. Ibid, September 1, 1842.

65. Richard B. Morris, ed., *Encyclopedia of American History* (New York: Harper & Row, 1970) 178-79.

66. AO, Canada Company Papers, *Letters from Widder*, Widder to Directors, September 25, 1842.

67. Ibid, October 11, 1842.

68. AO, Canada Company Papers, *Correspondence with Commissioners*, Perry to Commissioners, November 17, 1842.

69. The Corn Laws were passed in 1794 to protect British agriculture from outside competition. They were amended in the 1820s to give preference to colonial imports but were then repealed in 1846. In the short term, loss of preference in the British market was a blow for the colony, but by the prosperous 1850s the grain trade had recovered. See J.M.S. Careless, "Corn Laws," in *The Canadian Encyclopedia* (3 volumes) (Edmonton: Hurtig Publishers, 1985) 425.

70. AO, Canada Company Papers, *Correspondence with Commissioners*, Perry to Commissioners, August 11, 1842.

71. Ibid, December 29, 1842.

72. T.M. Jones to a local settler (name withheld), dated November 22, 1843. From a private collection of papers belonging to the family of the settler.

73. Toronto Public Libraries, Extract from *St. Catharines Journal* attached to Canada Company Broadside (1843).

74. AO, Canada Company Papers, *Correspondence with Commissioners*, Perry to Commissioners, February 2, 1843.

75. University of Guelph, Arts Division, *Report of the Court of Directors of the Canada Company to the Proprietors,* London, March 29, 1843.

76. AO, Canada Company Papers, *Minutes of the Court of Directors*, Committee of Correspondence, June 29, 1843. To put costs in Upper Canada into perspective, the Canada Company office in Toronto produced an information brochure dated June 5, 1843 for prospective settlers which included the following helpful information:

Price per Acre of the Company's Lands:
Huron District: 11s. 3p. to 15s. currency, per acre London, Brock and Talbot districts: 12s. 9d. to 17s 6d. currency per acre Bathurst, Eastern, Ottawa, and Dalhousie: 2 s. od. to 12s 6d. currency per acre

Wages: male farm servants per month with board: £2 (without board: £3); female servants: £1 per month with board; day labourers: 3s.9d. without board; carpenters and other tradesmen: 5s. to 10s.per day depending on ability.

Taxes: 1 1/2d. to £1 on assessed property

Representative costs: comfortable log house – two floors 16' by 24' with shingled roof: £9; frame house same size: £50; log barn: 24'x40': £10; frame barn same size: £70; household furniture: should not exceed £10; clothes: 50% addition over old country prices; crockery and cutlery: "very cheap."

Waggon: £15-20; double harness: £6-7 10s.; saddle and bridle: £3 15s; ploughs: £1 15s.; winnowing machines: £6 to £6 15s.; hay: £2 10s per ton.

Livestock: yoke of oxen: £10 to £12. 10s; cows: £3; farm horses: £10 to £15 (lower in some districts); sheep: 10s to 20s each.

Provisions: e.g., Pork:15s. to 20s per 100 lbs; flour £1 per barrel of 196 lbs; cheese: £1 10s. per cwt; butter £2 10s.; whiskey 1s to 1s. 3d. per gallon; beef: £1 to 1 5s. per cwt; oatmeal: 7s. per cwt."

Thelma Coleman, *The Canada Company, with supplement by James Anderson* (Stratford: County of Perth and Cumming Publishers, 1978) 117-120.

77. "The Canada Company," as reported in *London Morning Chronicle* and reprinted in the *British Colonist*, August 13, 1844.

78. Ibid.

79. Ibid.

80. W.H. Smith, *Canada: Past, Present and Future*, Vol. II (Toronto: publisher n.k., 1851) 164-166.

81. "The Canada Company," as reported in the *London Morning Chronicle* and reprinted in the *British Colonist*, May 3, 1844.

82. AO, Strachan Papers, *Strachan Letter Book 1844-1849*, Strachan to Canada Company Directors, October 6, 1844, 54.

83. "The Canada Company," as reported in the *London Morning Chronicle* and reprinted in the *British Colonist*, May 25, 1845.

84. AO, Merritt Papers, F. Widder to Rev. W.H. Ripley, November 18, 1844.

85. "The Clergy Reserves," *British Colonist*, December 20, 1844.

86. Ibid, December 31, 1844.

87. Bishop Strachan never publicly admitted to being the author of this work, but convincing proof is furnished in his letter book that the letters were in fact written by

him. The second paragraph of a letter to a Dr. Bethune, dated February 11, 1845, furnished this evidence: "I enclose my letters to Mr. Widder four in number. I have smoothed them down to a little more than a statement of facts and inferences. You will however get them copied in a fair hand & then when you can read them easily consider whether they can be printed and distributed. I should like them out before Mr. Jones returns. I must not have them traced to me which is the only difficulty. Were I an individual Clergyman or Layman I would adopt them without ceremony but my station forbids this." J.T [John Toronto] AO, Strachan Papers, Strachan Letter Book, (1844-1849) 75.

88. Toronto Public Libraries, Reference Division, "Aliquis," *Observations on the History and the Recent Proceedings of the Canada Company, addressed in Four Letters to Frederick Widder, Esq., one of the Commissioners* (Hamilton: 1845) 26.

89. Ibid, 27.

90. Ibid, 27-28. William Allan and John Strachan were close friends and so there was no way he would talk in disparaging terms about Allan who had left the company by this time in any case. They both had been members of the Executive Council of Upper Canada and Allan sat on the board of the (Anglican) Church Society which handled the monies for the Anglican Diocese of Toronto. Allan was very supportive of Bishop Strachan in the founding of Trinity College in 1851 and was a member of the board when he died in 1853.

91. Ibid, 31.

92. Ibid, 49.

93. AO, Canada Company Papers, *Minutes of Committees*, Committee of Correspondence, May 15, 1845.

94. AO, Canada Company Papers, *Correspondence with Commissioners*, Perry to Commissioners, May 19, 1845.

95. Ibid, May 3, 1845.

96. "The Canada Company," as reported in the *Montreal Gazette* and reprinted in the *British Colonist*, May 2, 1845.

97. For an account sympathetic to the Huron District Council, see Graham, 274-278.

98. See Merritt Papers, letter from F. Widder (probably to W.H. Merritt), January 13, 1845.

99. AO, Merritt Papers, *Memorandum for the Information of the Legislative Council of Canada*, not dated.

100. AO, Canada Company Papers, *Correspondence with Commissioners*, Perry to Commissioners, June 3, 1844.

101. AO, Merritt Papers, *Memorandum for the Information of the Legislative Council of Canada*, not dated.

102. William Hamilton Merritt is credited with the idea of building of the Welland Canal (it opened in 1829). He enlisted government support, raised the necessary funds and supervised the project. He was also the visionary behind a St. Lawrence canal system. Merritt was first elected to the Upper Canada Assembly in 1832, re-elected in 1834 and 1836, and after the union, elected in 1841, 1847, and 1848 to 1860. He first sat as

a Tory, then as a moderate Reformer. He was also a railway promoter and was very active as a legislator and businessman. "Merritt can be seen as one of the great figures in the history of Canadian transportation." See J.J. Talman, "Merritt, William Hamilton," in *Dictionary of Canadian Biography, Volume ix* (Toronto: University of Toronto Press, 1976) 544-48 for an overview of his life and work.

103. AO, Merritt Papers, F. Widder to William Hamilton Merritt, January 25, 1845.

104. Ibid, February 6, 1845.

105. Ibid.

106. "The Canada Company," as reported in the *Montreal Herald*, and reprinted in the *British Colonist*, May 13, 1845.

107. AO, Canada Company Papers, *Letters from Widder*, Widder to Directors, February 19, 1845.

108. Ibid.

109. Sales and leases dropped from 103,518 acres in 1845 to 48,627 acres in 1846, but rose again to 117,940 acres in 1847. From W.H. Smith, *Canada: Past, Present and Future*, Vol. ii (London: publisher n.k., 1851) 165.

110. AO, Canada Company Papers, *Correspondence with Commissioners*, Perry to Commissioners, July 3, 1845.

111. Ibid, May 3, 1845.

112. AO, Minutes of Committees, *Committee of Correspondence*, September 17, 1846.

113. Ibid, June 4, 1846.

114. Ibid, June 18, 1846.

115. Ibid, September 3, 1846.

116. Ibid, October 3, 1845.

117. T.M. Jones to a settler (name withheld) in the Huron Tract, December 1, 1846. From a private collection of papers belonging to the family of the settler.

118. AO, Canada Company Papers, *Minutes of Committees*, Committee of Correspondence, October 2, 1845.

119. In 1846, the estimated population of these communities was: Toronto – 19,706; Guelph – 1,240; Stratford – 200: Goderich – 659: Sarnia – 420; London – 3,500. From Wm. H. Smith, *Smith's Canadian Gazetteer* (Toronto: H. & W. Rowsell, 1846; Facsimile edition: Toronto: Coles, the Book people, 1970).

120. AO, Canada Company Papers, *Correspondence with Commissioners*, Perry to Commissioners, July 18, 1845.

121. Ibid, November 3, 1845.

122. Ibid, June 18, 1846.

123. For a detailed discussion of on the separation of the five townships and the development of the railways in the Huron Tract, see H.J. Johnston, "Transportation and the Development of the Eastern Section of the Huron Tract, 1828-1858," unpublished M.A. thesis, University of Western Ontario, 1965.

124. AO, Canada Company Papers, *Correspondence with Commissioners*, Perry to Commissioners, February 3, 1847.

125. Ibid, August 3, 1847.

126. University of Guelph, Canada Company Papers, see *Report of the Court of Directors of the Canada Company to the Proprietors* (London, March 28, 1849).

127. AO, Canada Company Papers, *Correspondence with Commissioners*, Perry to Commissioners, November 3, 1847.

128. University of Guelph, Arts Division, *Report of the Court of Directors of the Canada Company to the Proprietors* (London, March 21, 1850).

129. Mr. Robertson could well be the "Ross Robertson" referred to in the Lizars' book about an episode in the 1841 election. "Fights, is it? Oh, they don't amount to much; just a squabble at nights. If I'd a vote, I'd have polled it for Strachan. I was stopping at Caberfae's at the Gyard House, where some twelve of the First Royals were stopping. In Judge Acland's time there was a warrant out for seventeen of us, because it wasn't lagal for us to have walked into town. There was Moderwell, Ross Robertson, Bob Ellis, and some others who tried to stop us; but we wouldn't be stopped, so they took our seventeen names." Robina and Kathleen Macfarlane Lizars, *In The Days of the Canada Company,* 77.

130. AO, Canada Company Papers, *Correspondence with Commissioners*, Perry to Commissioners, September 6, 1850.

131. Ibid.

132. Ibid, Franks to Jones, February 21, 1851.

133. Ibid, Perry to Commissioners, June 27, 1851.

134. Ibid, October 31, 1851.

135. Ibid, January 16, 1852.

136. AO, Canada Company Papers, Letters from Frederick Widder to T.M. Jones, Widder to Jones, March 16, 1852.

137. Ibid, April 5, 1852.

138 A railway line from Toronto to Lake Huron via Guelph had been discussed and promoted as early as 1836 when a charter of incorporation was obtained, and then renewed in 1846, by the City of Toronto and Lake Huron Rail Road Company (CT&LHRRC). In 1851, the newly founded Toronto and Guelph Railroad Company (T&GRC) received its charter for a rail line from Toronto to Guelph, with the idea of extending it subsequently to Lake Huron.

    While Canada Company directors were busily promoting the Toronto-Guelph line as a first step, and then on to Goderich through Stratford, prominent Goderich citizen Daniel Lizars was hedging his bets by producing a pamphlet extolling the virtues of the town which could be used by the two Toronto groups as well as by proponents of the Buffalo-Brantford-Goderich line (which he publicly supported, although undoubtedly from his perspective, a line from Buffalo and/or a line from Toronto would be just fine).

    A terminus in Sarnia was attractive to promoters given its strategic location just across the border from the United States, at the base of Lake Huron where it empties into the St. Clair River. These promoters included the eleven member board of the CT&LHRRC which was now presided over by former Canada Company commissioner, William Allan, with Frederick Widder as a board member. Interestingly, Widder was also on the board of the Toronto and Guelph Railroad Company – and the Canada

Company agreed to act as agent in England for both railway companies in the disposal of their bonds on the London bond market. Obviously, the directors had changed their mind from the late 1830s about the threat Sarnia posed to Goderich and had decided to play "both ends against the middle." That said, while they did not want to be left out of the action, they still favoured Goderich. The Grand Trunk, though, was unbeatable and the CT&LHRR and the T&GRC never did get a chance to build their railways.

In the end, because of the success of the Buffalo and Lake Huron Railway promoters, the Buffalo-Brantford-Goderich line (via Stratford) was the first one built to Lake Huron. It opened officially in Goderich (with great fanfare) in June 1858. But there would be competition very soon. In 1857, the Grand Trunk Railway completed its main line from Toronto via Stratford to St. Marys. It had then built a secondary line from St. Marys to London the following year where it could connect with the Great Western Railway (this latter railway ran from Niagara Falls through Hamilton and London to Windsor). In 1858, the Grand Trunk then completed its main line by connecting St. Marys and Sarnia. Toronto and Sarnia were now linked. Any hope by Canada Company directors that Goderich would become an important entrepot on Lake Huron, given its water and rail connection, would not materialize.

But these were still heady days and the Buffalo and Lake Huron Railway (B&LHR) was full of optimism in 1859 when it purchased the Canada Company's 250 acre Goderich Harbour property for £13,000. Unfortunately the railway's enthusiasm was short-lived. The B&LHR would run into constant financial problems and five years later, in 1864, the railway would be leased to the Grand Trunk Railway which would subsequently take over ownership in 1870. In the meantime, the new management would build a grain elevator in 1866, but Goderich would continue to be a branch line terminus shipping primarily grain, salt and lumber – and later flour. The concerns of Canada Company directors expressed years earlier about the perceived competition from Sarnia would ring true. Sarnia had direct access to the United States and a better harbour which was not exposed to the storms from the northwest that made Goderich less attractive as a trans-shipment point.

The Guelph Junction Railway (GJR) – which used the Priory as its passenger station from 1887 to 1911 – and the industries of Guelph wanted a direct connection to a lake port and saw Goderich as the answer. The result was the Guelph & Goderich Railway Company which was chartered in 1904 by the GJR and which in turn reached an agreement with the Canadian Pacific Railway to construct and operate it. In 1907, the line from Guelph to Goderich was completed, but operation ceased in July 1988. The tracks were subsequently lifted and the line from Goderich to Carlow is now the Maitland Trail. The original GJR/CPR station at the harbour is the headquarters of the trail association.

The Canadian National Railway, which absorbed the bankrupt Grand Trunk Railway system in 1923, continued to operate into Goderich until 1992. That year, RailTex created the Goderich-Exeter Railway Company which took over ownership of the line from Stratford to Goderich. Much of the tonnage handled is salt as well as grain and heavy machinery. The original Buffalo and Lake Huron Railway passenger station in Goderich was split in two and moved to the bank of the lake on Essex Street in 1907. It became two summer cottages located on adjoining properties, one of which belonged to the author's family. In 1971, this latter building was moved along Essex Street to accommodate the construction of a new house and is now used as a year-round residence. A modern house also replaced the other cottage which was moved to another location in town in the mid-1960s. See Toronto Public Library (TRL) City of Toronto & Lake Huron Rail Road Co. Papers 1836-1847 *Letter Books – Letters of July 25 and July 26, 1845)*; James Marsh, "Grand Trunk Railway of Canada," and Peter Baskerville, "Great Western Railway," in *The Canadian Encyclopedia* (3 volumes) (Edmonton: Hurtig Publishers, 1985) 764 & 772. The following Web sites were also consulted, all in June 2004: "The Guelph Junction Railway" at http://www.globalserve.net/~robkath//railguju.htm; "The Guelph & Goderich Railway" at http://www.globalserve.net/~robkath/railgugo.htm; "Goderich-Exeter Railway" at http://www.trainweb.org/rosters/GEXR.html; and various *Guelph Advertiser* newspaper articles about the Toronto and Guelph Railway available at http://www.hhpl.on.ca/sigs/ehs/news.html (the Web site of the Esquesing Historical Society).

139. AO, Canada Company Papers, *Correspondence with Commissioners*, Perry to Commissioners, March 26, 1852.

140. Ibid, April 23, 1852.

141. AO, Canada Company Papers, *Letters from Frederick Widder to T.M. Jones*, Widder to Jones, April 20, 1852.

142. Although not generally known, this fact came to light on reviewing the Grand Trunk Railway web site, http://www.globalserve.net/~robkath/railgt.htm "Buffalo and Lake Huron Railway," downloaded May 2004. That John Galt Jr. became president is certainly within the realm of the possible given his interests and the people behind him in the community. Ties to railways ran in the family. His brother, Alexander Tilloch Galt, while working for the British American Land Company in Sherbrooke, Quebec, had promoted the concept of the St. Lawrence and Atlantic Railway. In 1849, he became president of that railway which was completed in 1853 and ran from Montreal to Portland, Maine.

143. AO, Canada Company Papers, *Correspondence with Commissioners*, Perry to Commissioners, July 23, 1852.

144. AO, Canada Company Papers, *Letters from Frederick Widder to T.M. Jones*, June 23, 1852.

145. LAC, Canada Company Papers, *Correspondence with Commissioners*, Perry to Commissioners, July 30, 1852.

146. Ibid.

147. Ibid.

148. Ibid, November 11, 1852.

149. Ibid, Perry to Jones, April 1, 1853.

150. For an overview of the life William B. Robinson, see Julia Jarvis, "Robinson, William Benjamin," in *Dictionary of Canadian Biography*, *Volume x* (Toronto: University of Toronto Press, 1972) 622-24.

151. LAC, Canada Company Papers, *Correspondence with Commissioners*, Perry to Commissioners, Perry to Widder, January 11, 1853.

152. On leaving the employ of the Canada Company in 1852, Thomas Mercer Jones and his wife left the Canada Company headquarters building in Goderich where they had lived for 12 years but stayed on in the town. They did not have far to go as to they moved across the street to a two-storey house on West Street which is still used as a residence. On September 16, 1853, Jones was appointed agent of the Bank of Montreal in Goderich. Elizabeth Mary's health suffered during this period and she died in Goderich in 1857 at the age of 43 (Jones was now 63).

The Lizars' account is touching: "General debility progressed into a definite complaint; the Ranee continued to fade; the Bishop and the equally heart-broken Commissioner watched; the light waned, went out, and improved times showed a note of crape at the door. It is said that some Scotch people from their religious side ... dearly love a 'judgment.' These people say that health has been undermined by a life of gaiety, and the tremulous thin-lipped Scotch mouth opined: 'Oo aye, she was juist a silly little addle-pate.' Not so. To please and to be easily pleased, to love, and to be loved, to live appreciated and die regretted, is the best of judgments on life. There was an exit as dramatic as had been the Tiger's last home-coming. A springless wagon in which rested, or rather did not rest, a coffin; a brother seated upon it; a departure in the dawn of an early dawn of a raw March morning, and the butterfly Queen followed, at a day's journey behind, the Commissioner and the Bishop. Travelling night and day, they took their way along the Huron Road. The sun tried to pierce the leaden skies ... But gloom came on again ... the belle, the toast of early Huron times, the chatelaine of the Canada Company passed out of sight." Robina and Kathleen Macfarlane Lizars, *In the Days of the Canada Company* (Toronto, William Briggs, 1896) 466-67.

Following Elizabeth Mary's death, Thomas Mercer Jones returned to Toronto where he lived in retirement. Archival records of St. James Cathedral, Toronto, indicate that he died in 1868, "from the effects of drink" according to his obituary; Charles Mercer returned to Goderich at some point and died there while Strachan Graham stayed in Toronto and died there in 1869. It is not known if the sons married or produced heirs.

153. LAC, Canada Company Papers, *Correspondence with Commissioners*, Perry to Commissioners, April 1, 1853.

154. Ibid, September 7, 1855.

155. LAC, Canada Company Papers, *Correspondence with Commissioners*, Franks to Jones, March 4, 1853.

Chapter 6: You Be the Judge

1. G.M. Craig, *Upper Canada: The Formative Years, 1784-1841* (Toronto: McClelland & Stewart, 1963) 138.

2. AO, Robinson Papers, Comments made in March 1853 (on a paper with related minutes) delivered by J.B. Robinson to R.J. Wilmot, January 10, 1823.

3. Ibid.

4. AO, Strachan Papers, *Strachan Letter Book, 1794-1891*, John Richardson to John Strachan, March 27, 1824.

5. "The Canada Company," as reported in the *Kingston Chronicle* and reprinted in the *British Colonist*, April 15, 1845.

Epilogue

1. Government of Ontario, Office of the Provincial Secretary, Canada Company File, 19 & 20 Victoria, Chapter 23, June, 1856.

2. University of Guelph, Arts Division, Canada Company Papers, *Report of the Court of Directors of the Canada Company to the Proprietors,* London, March 21, 1878.

3. Ibid.

4. Thelma Coleman, *The Canada Company, with supplement by James Anderson* (Stratford: County of Perth and Cumming Publishers, 1978) 148.

5. Ibid.

6. Government of Ontario, Office of the Provincial Secretary, Canada Company file, The Canada Company's Act, 1916,

7. Ibid.

8. University of Guelph, Arts Division, Canada Company Papers, *Report of the Court of Directors of the Canada Company to the Proprietors*, London, May 17, 1939.

9. Government of Ontario, Office of the Provincial Secretary, Canada Company File, The Companies Information Act, Return to the Provincial Secretary, Ontario: Information and Particulars as of May 1, 1938.

10. Ibid, March 31, 1948.

11. Ibid, March 31, 1949.

12. Reuben R. Sallows was born in Huron County near Goderich in 1855. At age 21, he apprenticed as a photographer. In 1897, he went from being a studio photographer to launching a career as one of the most important photographers of his generation, featuring domestic scenes, rural life, nature and various outdoor pastimes.

13. Government of Ontario, Information and Particulars as of May 1, 1938.

14. Government of Ontario, Office of the Provincial Secretary, Canada Company File: Smallfield, Rawlins and Company, London, England, to the Office of the Deputy Provincial Secretary, Toronto, October 28, 1953.

15. Ibid, September 4, 1954.

16. Government of Ontario, Office of the Provincial Secretary, Canada Company File: Order Cancelling Extra-Provincial License, September 11, 1961.

Appendix B: Canada Company Directors

1. AO, Canada Company Papers, *Minutes of the Courts of Directors*, May 14, 1824; July 30, 1824; August 24, 1826; February 19, 1829; University of Guelph, Canada Company Papers, *Charter of Incorporation, August 19, 1826.*

Appendix C: Huron Tract Township Names and Their Origins

Sources:

Dictionary of Canadian Biography. Toronto: University of Toronto Press, 1991.

Dictionary of National Biography: From the Earliest Times to 1900. London: Oxford University Press, published since 1917.

Coleman, Thelma, *The Canada Company (With Supplement by James Anderson)*. Stratford: County of Perth and Cummings Publishers, 1978.

AO, Canada Company *Minutes of the Court of Directors*.

Appendix D: "The Huron Tract" by Dr. William "Tiger" Dunlop, 1841

1.   AO, F129 B-3 Vol. 1. TRHT Intro A06473, Description of Huron Tract Lands by Dunlop 1841, (Introduction to CanCo Atlas, with text reproduced verbatim to show perception of the Tract as of 1841).

BIBLIOGRAPHY

I. BOOKS

a) Contemporary Accounts

Bonnycastle, Sir Richard H., *Canada and the Canadians*. London: H. Colburn, 1849.

_____, *The Canadas in 1841*. London, H. Colburn: 1841.

Brown J.B., *Views of Canada and the Colonists*. Edinburgh: 1851

Carruthers, J, *Retrospect of Thirty-Six Years' Residence in Canada West*. Hamilton: 1861.

Doyle, Martin (pseudonym), *Hints on Emigration to Upper Canada*. Dublin: 1831.

Dunlop, W., *Statistical Sketches of Upper Canada for the Use of Emigrants*. London: 1832.

Durham, John George Lambton, Earl of, *The Report of the Earl of Durham*. London: Methuen, 1902.

Fergusson, Adam, *Practical Notes Made During a Tour in Canada, and a Portion of the United States*. Edinburgh: 1834.

Fidler, Rev. Isaac, *Observations of Professions, Literature, Manners and Emigration in the United States and Canada Made During a Residence There*. New York: 1833.

Galt, John, *The Autobiography of John Galt*. 2 vols. London: Cochrane and M'Crone, 1833.

_____, *Lawrie Todd, or, The Settlers in the Woods*. London: R. Bentley, 1832.

Gillespie, W.M, *A Manual of the Principles and Practice of Road-Making, Comprising the Location, Construction and Improvement of Roads and Rail-Roads*. New York: 1855.

Gourlay, Robert, *General Introduction to Statistical Account of Upper Canada*. New York: Johnson Reprint Corporation, 1966.

Harrison, J., *Sketches of Upper Canada*. Edinburgh: 1825.

Jameson, Anna Brownell, *Winter Studies and Summer Rambles in Canada*. 2 vols. New York: Wiley and Putnam, 1839.

Mackenzie, William Lyon, *The Selected Writings of William Lyon Mackenzie, 1824-1837.* Edited by Margaret Fairley. Toronto: Oxford University Press, 1960.

Magrath, T.W., *Authentic Letters from Upper Canada.* Edited by Rev. T. Radcliff. Dublin: 1833.

Rolph, Thomas, *Emigration and Colonization.* London: 1844.

Statistical Account of Upper Canada. Dundas, 1836.

Smith, Wm. H., *Canada: Past, Present and Future.* 2 vols. Toronto, 1861.

_____, *Smith's Canadian Gazetteer.* Toronto: H. & W. Rowsell, 1846; Facsimile edition, Toronto: Coles, the Book People!, 1970.

Strickland, Samuel, *Twenty-Seven Years in Canada West, or, The Experience of an Early Settler.* 2 vols. London: R. Bentley, 1853.

Traill, Catharine Parr, *The Backwoods of Canada.* London: C. Knight, 1836, Reprinted Toronto: McClelland & Stewart, 1929.

Widder, F., *Information for Intending Emigrants of all Classes to Upper Canada.* Toronto, 1855.

b) Secondary Accounts

Aberdein, Jennie W., *John Galt.* London: Oxford University Press, 1963.

Armstrong, G.H., *The Origin and Meaning of Place Names in Canada.* Toronto: Macmillan, 1930.

Bethune, A.N., *Memoir of The Right Reverend Bishop Strachan.* Toronto: Henry Rowsell, 1870.

Bliss, J.M., *Canadian History in Documents, 1763-1966.* Toronto: Ryerson, 1966.

Brown, G., "The Canada Company." Unpublished MBA thesis, University of Western Ontario, 1936.

Burrows, C.A., *The Annals of the Town of Guelph.* Guelph: Harold Steam Printing House, 1877.

Campbell, Wilfred, *The Scotsman in Canada.* Vol. I. Toronto: Musson, n.d.

Campey, Lucille H., *"A Very Fine Class of Immigrants": Prince Edward Island's Scottish Pioneers, 1770-1850.* Toronto: Natural Heritage, 2001.

_____, *After the Hector: The Scottish Pioneers of Nova Scotia and Cape Breton, 1773-1852.* Toronto: Natural Heritage, 2004.

_____, *"Fast Sailing and Copper-Bottomed": Aberdeen Sailing Ships and the Emigrant Scots They Carried to Canada, 1774-1855.* Toronto: Natural Heritage, 2002.

_____, *The Silver Chief: Lord Selkirk and the Scottish Pioneers of Belfast, Baldoon and Red River.* Toronto: Natural Heritage, 2003.

Canniff, William, *History of the Settlement of Upper Canada.* Toronto: Dudley and Burns, 1869.

"Canuck, A.," *Pen Pictures of Early Pioneer Life in Upper Canada.* Toronto: William Briggs, 1905.

Carrothers, W.A., *Emigration from the British Isles.* London: P.S. King and Son, 1929.

Clarke, Charles, *Sixty Years in Upper Canada.* Toronto: William Briggs, 1908.

Connant, Thomas, *Life in Canada.* Toronto: William Briggs, 1903.

Connon, John R., *The Early History of Elora, Ontario, and Vicinity*. Elora: The Elora Express and The Fergus News Record, 1930.

Coleman, Thelma, with supplement by James Anderson, *The Canada Company*. Stratford: County of Perth & Cummings Publishers, 1978.

Craig, Gerald M., ed., *Early Travellers in Upper Canada, 1791-1867*. Toronto: Macmillan, 1955.

_____, *Upper Canada: The Formative Years, 1784-1841*. Toronto: McClelland & Stewart, 1963.

Creighton, D.G., *The Commercial Empire of the St. Lawrence, 1760-1850*. Toronto: Ryerson, 1937.

Day, Frank, *Here and There in Eramosa*. Guelph: Leaman Printing Co., 1963.

Dobbin, F.H., *Our Old Home Town*. Toronto: Dent, 1943.

Dunham, Aileen, *Political Unrest in Upper Canada*. Introduction by A.L., Burt. The Carleton Library Series #10. Toronto: McClelland & Stewart, 1963.

Dunham, Mabel, *Grand River*. Toronto: McClelland & Stewart, 1945.

Ermatinger, C.O., *The Talbot Regime, or, The First Half Century of the Talbot Settlement*. St. Thomas, Ontario: Municipal World, 1904.

Ford, F.S.L., *William Dunlop, 1792-1848*. 2nd Edition. Toronto: The Albert Britnell Book Shop, 1934.

Fox, W.S., *T'Ain't Runnin' No More – Twenty Years After*. London: Oxford Book Shop Ltd., 1958.

Fraser, Alexander, *Sixteenth Report of the Department of Archives for the Province of Ontario*. Toronto: The King's Printer, 1920.

Gates, Lillian & F. Gates, *Land Policies of Upper Canada*, Toronto: University of Toronto Press, 1968.

Gazetteer and Directory of the County of Perth. Toronto: H. Beldon and Co., 1879.

Gentilcore, R. Louis & C. Grant Head, *Ontario's History in Maps*. Toronto: University of Toronto Press, 1984.

Gibbon, J.M., *Scots in Canada*. Toronto: Musson, 1911.

Gillis, D.H., *Democracy in the Canadas, 1759-1867*. Toronto: Oxford University Press, 1951.

Glazebrook, G.P. de T., *Life in Ontario: A Social History*. Toronto: University of Toronto Press, 1968.

Graham, W.H., *The Tiger of Canada West*. Toronto: Clarke, Irwin, 1962.

Grant, George Munro, ed., *Picturesque Canada: The Century as it Was and Is*. Toronto: Beldon Bros., 1882.

Gregg, William, *History of the Presbyterian Church in Canada*. Toronto: Presbyterian Printing and Publishing Co., 1885.

Guillet, Edwin C. *Early Life in Canada*. Toronto: The Ontario Publishing Co. Ltd., 1933.

_____, *The Pioneer Farmer and Backwoodsman*. 2 vols. Toronto: The Ontario Publishing Co. Ltd., 1963.

Haight, Canniff, *Country Life in Canada Fifty Years Ago*. Toronto: Hunter, Rose & Co., 1885.

Hathaway, E.J., *Jesse Ketchum and His Time*. Toronto: McClelland & Stewart, 1929.

Hind, H.Y., and others., *The Dominion of Canada*. Toronto: L. Stebbins, 1868.

Historical Atlas of the County of Wellington. Toronto: H. Beldon and Co., 1879.

Hodgins, J. George, *A History of Canada*. Montreal: John Lovell, 1866.

Huron County Historical Society, *Huron Historical Notes*. Vols. I and II. **[dates?]**

Illustrated Historical Atlas of the County of Huron. Toronto: H. Beldon and Co., 1879.

Illustrated Historical Atlas of the County of Perth. Toronto: H. Beldon and Co., 1879.

Johnson, Stanley C., *A Historical Emigration from the United Kingdom to North America, 1763-1912*. London: George Routledge and Sons Ltd., 1913.

Johnston, William, *History of the County of Perth, 1825-1902*. Stratford: At the Beacon Office, 1903.

_____, *The Pioneers of Blanshard: With an Historical Sketch of the Township*. Toronto: William Briggs, 1899.

Kennedy, David, *Incidents of Pioneer Days at Guelph and the County of Bruce*. Toronto: Department of Agriculture, 1903.

Kerr, D.G.G., and R.J.K. Davidson, *Canada: A Visual History*. Toronto: Nelson, 1966.

Kingsford, William, *The History of Canada*. Toronto: Rowsell & Hutchinson, 1898.

Klinck, Carl F., *William "Tiger" Dunlop*. Toronto: Ryerson, 1958.

Landon, Fred, *Lake Huron*. New York: The Bobbs-Merrill Co., 1944.

_____, *Western Ontario and the American Frontier*. Toronto: Ryerson, 1941.

Langton, W.A., *Early Days in Upper Canada: The Letters of John Langton*. Toronto: Macmillan, 1926.

Lizars, Robina and Kathleen M., *In the Days of the Canada Company*. Toronto: William Briggs, 1896.

Macdonald, Norman, *Immigration and Settlement. The Administration of the Imperial Land Regulations*. Aberdeen: Longman's Green, 1939.

McDougall, Robert, *The Emigrant's Guide to North America*. Edited by Elizabeth Thompson. Toronto: Natural Heritage, 1998.

McInnis, Edgar, *Canada: A Political and Social History*. Toronto: Holt, Rinehart & Winston, 1963.

Meredith, Alden G., *Mary's Rosedale and Gossip of Little York*. Ottawa: The Graphic Publishers Ltd., 1928.

Miller, Orlo, *Middlesex County*. London: Civic Sales and Services, 1964.

Morgan, Henry J., *Sketches of Celebrated Canadians*. Quebec: Hunter, Rose and Co., 1862.

Morrison, Jean, *Superior Rendezvous-Place: Fort William in the Canadian Fur Trade*. Toronto: Natural Heritage, 2001.

Morton, Desmond, *A Short History of Canada*. Edmonton: Hurtig Publishers, 1983.

Murray, Florence B., ed., *Muskoka and Haliburton, 1615-1876*. Toronto: The Champlain Society, 1963.

Needler, G.H., *Colonel Anthony Van Egmond: From Napoleon and Waterloo to Mackenzie and Rebellion*. Toronto: Burns and MacEachern, 1956.

_____, *Otonabee Pioneers: The Story of the Stewarts, the Stricklands, the Traills and the Moodies*. Toronto: Burns and MacEachern, 1953.

North, Anison, *The Forging of the Pikes*. Toronto: Copp Clark, 1942.

Read, D.B., *The Lieutenant-Governors of Upper Canada and Ontario, 1792-1899*. Toronto: William Briggs, 1900.

Reynolds, Lloyd G., *The British Immigrant*. Toronto: Oxford University Press, 1935.

Robertson, J. Ross., *Robertson's Landmarks of Toronto*. Toronto: J. Ross Robertson, 1894.

Robinson, C.W., *Life of Sir John Beverley Robinson*. Toronto: Morang and Co. Ltd., 1904.

Scadding, Henry, *Toronto of Old*. Toronto: Adam Stevenson and Co., 1873.

Scott, James, *The Settlement of Huron County*. Toronto: The Ryerson Press, 1966.

Shortt, Adam, & Arthur G. Doughty, *Canada and Its Provinces*. 23 volumes. Toronto: Glasgow, Brook and Co., 1914-1917.

Skelton, Oscar Douglas, *Life and Times of Sir Alexander Tilloch Galt*. Toronto: Oxford University Press, 1920.

Stephen, Sir Leslie, & Sir Sidney Lee, eds., *Dictionary of National Biography: From the Earliest Times to 1900*. London: Oxford University Press, published since 1917.

Thompson, Samuel, *Reminiscences of a Canadian Pioneer for the Last Fifty Years*. Toronto: Hunter, Rose and Co., 1884.

Waite, P.B., ed., *Pre-Confederation*. Vol. II. Canadian Historical Documents Series. Scarborough: Prentice-Hall, 1965.

Wallace, W. Stewart, *The Dictionary of Canadian Biography*. Toronto: Macmillan, 1928.

Weaver, Emily P., *The Story of the Counties of Ontario*. Toronto: Bell and Cockburn, 1913.

Withrow, William H., *A Popular History of the Dominion of Canada*. Toronto: William Briggs, 1884.

Wittle, Carl, *A History of Canada*. Toronto: McClelland & Stewart, 1941.

Wood, H.F., *Forgotten Canadians*. Toronto: Longmans, 1963.

Wrong, George M. & H.H. Langton, eds., *The Family Compact. Part VII. The Struggle for Political Freedom*. Chronicles of Canada. Toronto: Glasgow, Brook and Co., 1915.

Young, James, *Reminiscences of the Early History of Galt and the Settlement of Dumfries in the Province of Ontario*. Toronto: Hunter, Rose and Co., 1880.

II. NEWSPAPERS

Goderich, Ontario – *Signal-Star*.

Hamilton, Ontario – Western *Mercury*.

Kingston, Ontario – *Kingston Chronicle*.

Niagara Falls, Ontario – *Niagara Gleaner*.

Toronto, Ontario – *British Colonist; Colonial Advocate; Correspondent and Advocate*;

Toronto, Ontario – *Patriot and Farmer's Monitor; Upper Canada Herald*.

III. OTHER ARTICLES AND PAPERS

Cameron, James M., "Guelph and the Canada Company, 1827-1851: An Approach to Resource Development." Unpublished M.Sc. thesis, University of Guelph, 1966.

Gordon, R.K., "John Galt," University of Toronto Philology #5. Toronto, 1920.

Johnston, H.J., "Immigration to the Five Eastern Townships of the Huron Tract," *Ontario Historical Society Papers and Records*. Vol. LIV. Toronto: 1962.

Johnston, H.J., "Transportation and the Development of the Eastern Section of the Huron Tract, 1828-1858." Unpublished M.A. thesis, University of Western Ontario, 1965.

Karr, C.G., "The Foundations of the Canada Land Company, 1823-1843." Unpublished M.A. thesis, University of Western Ontario, 1966.

Lewis, Paul Eugene, "Faction in the Goderich Area, Huron County, 1835-1841." Unpublished M.A. thesis, University of Western Ontario, 1967.

Wilson, G.A., "The Political and Administrative History of the Upper Canada Clergy Reserves, 1790-1855," Ph.D. thesis, University of Toronto, 1959.

IV. PRIMARY SOURCES

1. Archives of Ontario

a) Canada Company Papers

Agreements with His Majesty's Government and Act of Parliament.

Canada Company Letterbooks, February 9, 1824-March 31, 1831; March 10, 1834-January 20, 1848.

Commissioners' Letters and Reports, Vol. I-November 23, 1826-December 31, 1828; Vol. II-January 2, 1829-May 8, 1834.

Correspondence with Commissioners, Vol. I-May 22, 1834-October 18, 1842; Vol. II-November 3, 1842-March 22, 150; Vol. III-April 5, 1840-September 26, 1856.

Correspondence with His Majesty's Government, August 24, 1826-November 11, 1842.

Correspondence with Shareholders, May 14, 1831-December 2, 1831.

Letters to the Court of Directors from Frederick Widder, Vol. I-October 12, 1839-July 25, 1845 Vol. II-March 13, 1852-November 18, 1852; Vol. III-January 7, 1853-December 31, 1859.

Letters from Frederick Widder to T.M. Jones, March 13, 1852-November 18, 1852.

Letters from Frederick Widder to W.B. Robinson, March 1, 1853-June 8, 1853.

Minutes of the Committees, Vol. I-August 17, 1824-July 28, 1826; Vol. II-August 31, 1826-December 31 1829; Vol. III-April 19, 1827-August 1, 127; Vol. IV-January 5, 1830-January 23, 1834; Vol. V-February 6, 1834-September 13, 1838; Vol. VI-September 20, 1838-May 3, 1843; Vol. VII-May 11, 1843-November 11, 1847; Vol. VIII-November 18, 1847-August 12, 1869.

Minutes of the Court of Directors, Vol. I-July 30, 1824-November 17, 1826; Vol. II-August 24, 1826-July 26, 1832; Vol. III-August 30, 1832-December 31 1840; Vol. IV-July 8, 1841-January 20, 1848.

Other Letters of the Board of Directors, March 10, 1834-January 20, 1848.

Proceedings of the General Court, Vol. I-December 20, 1826-December 21, 1854.

Reports of the Court of Directors of the Canada Company to the Proprietors: 1839-1845.

b) Others – Canada Company

1839-Report, Affairs of the Canada Company to His Excellency the Right Honourable C. Poulett Thomson by Daniel Lizars.

1840-Miscellaneous Documents Relating to the Commission of Enquiry into the Affairs of the Canada Company in 1840.

1848-Catechism of Information for Intending Emigrants.
Other Crown Land Papers, Correspondence.
Scrapbook. *"The Canada Company and A. Van Egmond: The Story of 1837 in Huron County,"* by W.B. Kerr, 1940.

c) Special Collections

Baldwin Family Papers.
Cartwright Family Papers.
A.T. Galt Papers.
Eamilius E. Irving Papers.
Macaulay Family Papers.
William Hamilton Merritt Papers.
O'Brien Journals.
John Beverley Robinson Papers.
John Strachan Papers.
John Strachan Letter Books.
Samuel Street Papers.

2. Library and Archives Canada

Canada Company Papers: 1825-1887 (M.G. 24, I. 46).
Colonial Office Records, Q Series.
Executive Council Records (R.G. 1).
Miscellaneous Records Relating to the Canada Company (R.G. 4, B. 47).
Proceedings of the Executive Council Relating to the Canada Company (R.G. 1, L. 7, Vols. XLI-XLIII).
Report of Settlers Placed on Crown Lands by the Canada Company (R.G. 1, L. 7, Vol. XLIV).
Upper Canada Journals of the Legislative Assembly, 1818-1821. Upper Canada Land Petitions (R.G. 1, L. 3).
Ontario: Office of the Provincial Secretary, Canada Company File 19 & 20 Victoria, c. 23.
The Canada Company's Amendment Act, 1881.
Petition for Licence for Extra-Provincial Corporation.
Statement of Affairs of the Canada Company, December 31, 1901.
Statement of Affairs of the Canada Company, December 31, 1906.
The Canada Company's Act, 1916. Returns–under the Companies Information Act.

3. Toronto Public Library (TRL)

Act to Grant Lands in Upper Canada, 1825.
Act to Amend Act to Grant Lands in Province of Upper Canada, 1828.
"Aliquis" (pseudonym). *Observations on the History and Recent Proceedings of the Canada Company*, Hamilton, 1854.

A Statement of the Satisfactory Results which have Attended Emigration to Upper Canada from the Establishment of the Canada Company until the Present Period. London, 1841.

A Warning to the Canadian Land Company in a Letter Addressed to that Body by an Englishman Resident in Upper Canada. Kingston, 1824.

City of Toronto & Lake Huron Rail Road Co. Papers 1836–1847.

Canada Company. *Accounts of all Monies Paid and Payable by the Canada Company.* March 10, 1831.

Canada Company Advertising Pamphlets.

Canada Company Broadsides. Canada Company Prospectus.

Letters from Settlers in the Huron District. London, 1842.

Statements by Settlers on Canada Company Land in the Huron District (collected for the company by J.J.E. Linton). London, 1842.

Summary of Information and Evidence Relative to the Canadian Company. London, 1824.

4. University of Guelph

a) Canada Company Papers

Aitchison Brown Diary (microfilm copy).

Anonymous: *To the Shareholders of the Canada Company.* London, 1832.

Canada Company advertising material.

Charter of Incorporation, August 19, 1826. London, 1832. Information Relative to the Canadian Company. N.d.

Letters and Extracts of Letters from Settlers in Upper Canada. London, 183+–.

Reports of the Court of Directors of the Canada Company to the Proprietors: 1827–1835, 1878.

Smith, Thomas. "*An Address to the Shareholders of the Canada Company,*" London, 1829.

5. University of Western Ontario, Regional History Collection

John Galt Jr. Letter Book.

Henry Ransford Diary.

Bruce Smith Papers.

INDEX

ABOUT THE AUTHOR

Bob Lee, diplomat and historian of Scottish, English and Welsh heritage, was born and educated in Toronto but spent all of his summers as a boy in Goderich where his Huron County roots run deep. His great-great-grandfather Charles George Middleton arrived in Goderich Township from England (via Toronto in 1834 with his young wife, Elizabeth, and their first-born son) where he initially purchased 80 acres from the Canada Company and built a log cabin there. His great-grandfather, William Lee, arrived in Goderich in the 1850s and served as mayor in 1869 (he was a marine agent and ship chandler at the Goderich harbour). The author's grandfather, Charles Crabb Lee was mayor of Goderich (1931-34) and proprietor of a ship chandlery/coal/plumbing/hardware business at the harbour. He also owned and operated a summer resort in Goderich, the Sunset Hotel, until his death in 1944.

Bob studied at Bishop's University Lennoxville, Quebec (Hons. B.A. History/Political Science), under eminent Canadian historian Dr. Donald C. Masters. He subsequently went on to the University of Guelph at Guelph, Ontario, for his Masters degree. Under the tutelage of Dr. Masters, who had since moved to Guelph, he wrote his thesis "The Canada Company: A Study in Direction, 1826-1853." In 1967, Bob joined the Canadian Foreign Service as a Trade Commissioner and has had postings in the United States, Yugoslavia, Japan, Korea and Indonesia. His latest assignment in the Department of Foreign Affairs and International Trade, Ottawa, was as director of the department's Science and Technology (S&T) Program.

He lives in Ottawa with his wife, Young-Hae, who is president of the Canada–Korea Society. They have three children – Geoffrey, Jennifer and Stephen.